T0254942

Lecture Notes in Computer Science 13596

More information about this series at https://link.springer.com/bookseries/558

Ahmed Abdulkadir · Deepti R. Bathula · Nicha C. Dvornek · Mohamad Habes · Seyed Mostafa Kia · Vinod Kumar · Thomas Wolfers (Eds.)

Machine Learning in Clinical Neuroimaging

5th International Workshop, MLCN 2022 Held in Conjunction with MICCAI 2022 Singapore, September 18, 2022 Proceedings

Editors
Ahmed Abdulkadir
Lausanne University Hospital
Lausanne, Switzerland

Nicha C. Dvornek
Yale University
New Haven, CT, USA

Seyed Mostafa Kia
Donders Institute
Nijmegen, The Netherlands

Thomas Wolfers
University of Tübingen
Tübingen, Germany

Deepti R. Bathula
Indian Institute of Technology Ropar
Rupnagar, India

Mohamad Habes
The University of Texas Health Science Center
San Antonio, TX, USA

Vinod Kumar
Max Planck Institute for Biological Cybernetics
Tübingen, Germany

ISSN 0302-9743 ISSN 1611-3349 (electronic)
Lecture Notes in Computer Science
ISBN 978-3-031-17898-6 ISBN 978-3-031-17899-3 (eBook)
https://doi.org/10.1007/978-3-031-17899-3

This Springer imprint is published by the registered company Springer Nature Switzerland AG
The registered company address is: Gewerbestrasse 11, 6330 Cham, Switzerland

Preface

The increasing complexity and availability of neuroimaging data, computational resources, and algorithms have the potential to exponentially accelerate discoveries in the field of clinical neuroscience. However, while computational methods have become increasingly complex, the size and diversity of typical evaluation data sets has not increased at the same rate. As a result, findings may not generalize to the general population or be biased towards majorities. In our view, integrating knowledge across domains is an effective and responsible way to reduce the translational gap between methodological innovations, clinical research, and, eventually, clinical application. With this workshop, we established a platform for the exchange of ideas between engineers and clinicians in neuroimaging.

The 5th International Workshop on Machine Learning in Clinical Neuroimaging (MLCN 2022) was held on September 18, 2022, as a satellite event of the 25th International Conference on Medical Imaging Computing and Computer Assisted Intervention (MICCAI 2022) in Singapore, to continue the yearly recurring dialogue between experts in machine learning and clinical neuroimaging. The call for papers was made on May 2, 2022, and the submission period closed on July 8, 2022. Each submitted manuscript was reviewed by three or more Program Committee members in a double-blind review process.

The accepted papers were methodologically sound, thematically fitting, and contained novel contributions to the field of clinical neuroimaging; these were presented and discussed at the hybrid MLCN 2022 workshop by one of the authors. The contributions covered a wide range of in vivo image analysis for clinical neuroscience and were classified according to their focus as either Morphometry or Diagnostics, Aging, and Neurodegeneration. Morphometry of anatomical regions of interest or pathological tissue was investigated from multiple angles. Among those papers concerned with Morphometry, some methods for improving the segmentation accuracy or volumetry were proposed. Others investigated cross-modality consistency of volumetry or potential lower accuracy of segmentation in minority populations caused by imbalanced training data. The papers concerned with Diagnostics, Aging, and Neurodegeneration presented predictive machine learning models, statistical group comparisons, or modeling of trajectories. The applications included age prediction, dementia, atrophy, autism spectrum disorder, and traumatic brain injury.

Once more, this workshop was put together by a dedicated community of authors, committee members, and workshop participants. We thank all presenters and attendees for their valuable contributions that made the MLCN 2022 workshop a success.

September 2022

Ahmed Abdulkadir
Deepti Reddy Bathula
Nicha Chitphakdithai Dvornek
Mohamad Habes
Seyed Mostafa Kia
Vinod Jangir Kumar
Thomas Wolfers

Organization

Steering Committee

Christos Davatzikos	University of Pennsylvania, USA
Andre Marquand	Radboudumc, The Netherlands
Jonas Richiardi	Lausanne University Hospital and University of Lausanne, Switzerland
Emma Robinson	King's College London, UK

Organizing Committee/Program Committee Chairs

Ahmed Abdulkadir	Lausanne University Hospital, Lausanne, Switzerland
Deepti Reddy Bathula	Indian Institute of Technology Ropar, India
Nicha Chitphakdithai Dvornek	Yale University, USA
Mohamad Habes	University of Texas Health Science Center at San Antonio, USA
Seyed Mostafa Kia	University Medical Center Utrecht, The Netherlands
Vinod Jangir Kumar	Max Planck Institute for Biological Cybernetics, Germany
Thomas Wolfers	University Clinic Tübingen, Germany

Program Committee

Anoop Benet Nirmala	University of Texas Health Science Center at San Antonio, USA
Özgün Çiçek	CYTENA, Germany
Onat Dalmaz	Bilkent Üniversitesi, Turkey
Augustijn de Boer	Radboud University, The Netherlands
Niharika S. D'Souza	IBM, USA
Edouard Duchesnay	NeuroSpin, France
Benoit Dufumier	NeuroSpin, France
Martin Dyrba	German Center for Neurodegenerative Diseases, Germany
Ahmed Elgazzar	Academisch Medisch Centrum Universiteit van Amsterdam, The Netherlands
Xueqi Guo	Yale University, USA

Francesco La Rosa	Ecole polytechnique fédérale de Lausanne, Switzerland
Sarah Lee	Amallis Consulting, UK
Esten Leonardsen	University of Oslo, Norway
Tommy Löfstedt	Umeå universitet, Sweden
John A. Onofrey	Yale University, USA
Saige Rutherford	Radboud University Medical Center, The Netherlands
Anil Kumar Sao	IIT Mandi, India
Apoorva Sikka	Contextflow, Austria
Sourena Soheili Nezhad	Radboud University Medical Center, The Netherlands
Lawrence Staib	Yale University, USA
Didem Stark	Bernstein Center for Computational Neuroscience, Germany
Archana Venkataraman	Johns Hopkins University, USA
Matthias Wilms	University of Calgary, Canada
Tianbo Xu	University College London, UK
Xing Yao	Vanderbilt University, USA
Mariam Zabihi	Radboud University Medical Center, The Netherlands
Yuan Zhou	Fudan University, China
Juntang Zhuang	OpenAI, USA

Contents

Morphometry

Joint Reconstruction and Parcellation of Cortical Surfaces

Anne-Marie Rickmann[1,2(✉)], Fabian Bongratz[2], Sebastian Pölsterl[1], Ignacio Sarasua[1,2], and Christian Wachinger[1,2]

[1] Ludwig-Maximilians-University, Munich, Germany
[2] Lab for Artificial Intelligence in Medical Imaging,Technical University of Munich, Munich, Germany
arickman@med.lmu.de

Abstract. The reconstruction of cerebral cortex surfaces from brain MRI scans is instrumental for the analysis of brain morphology and the detection of cortical thinning in neurodegenerative diseases like Alzheimer's disease (AD). Moreover, for a fine-grained analysis of atrophy patterns, the parcellation of the cortical surfaces into individual brain regions is required. For the former task, powerful deep learning approaches, which provide highly accurate brain surfaces of tissue boundaries from input MRI scans in seconds, have recently been proposed. However, these methods do not come with the ability to provide a parcellation of the reconstructed surfaces. Instead, separate brain-parcellation methods have been developed, which typically consider the cortical surfaces as given, often computed beforehand with FreeSurfer. In this work, we propose two options, one based on a graph classification branch and another based on a novel generic 3D reconstruction loss, to augment template-deformation algorithms such that the surface meshes directly come with an atlas-based brain parcellation. By combining both options with two of the latest cortical surface reconstruction algorithms, we attain highly accurate parcellations with a Dice score of 90.2 (graph classification branch) and 90.4 (novel reconstruction loss) together with state-of-the-art surfaces.

1 Introduction

The reconstruction of cerebral cortex surfaces from brain MRI scans remains an important task for the analysis of brain morphology and the detection of cortical thinning in neurodegenerative diseases like Alzheimer's disease (AD) [28]. Moreover, an accurate parcellation of the cortex into distinct regions is essential to understand its inner working principles as it facilitates the location and the comparison of measurements [9,13]. While voxel-based segmentations are useful for volumetric measurements of subcortical structures, they are merely

A.-M. Rickmann and F. Bongratz—Equal contribution.

Supplementary Information The online version contains supplementary material available at https://doi.org/10.1007/978-3-031-17899-3_1.

A. Abdulkadir et al. (Eds.): MLCN 2022, LNCS 13596, pp. 3–12, 2022.
https://doi.org/10.1007/978-3-031-17899-3_1

suited to represent the tightly folded and thin (thickness in the range of few millimeters [24]) geometry of the cerebral cortex.

The traditional software pipeline FreeSurfer [10], which is commonly used in brain research, addresses this issue by offering a surface-based analysis in addition to the voxel-based image processing stream. More precisely, the voxel stream provides a voxel-based segmentation of the cortex and subcortical structures, whereas the surface-based stream creates cortical surfaces and a cortex parcellation on the vertex level. To this end, FreeSurfer registers the surfaces to a spherical atlas. Cortical thickness can be computed from these surfaces with sub-millimeter accuracy and different regions of the brain can easily be analyzed given the cortex parcellation. Yet, the applicability of FreeSurfer is limited by its lengthy runtime (multiple hours per brain scan).

Recently, significantly faster deep learning-based approaches for cortical surface reconstruction have been proposed [1,4,20,23]; they reconstruct cortical surfaces from an MRI scan within seconds. To date, however, these methods do not come with the ability to provide a parcellation of the surfaces. At the same time, recent parcellation methods [5,14] usually rely on FreeSurfer for the extraction of the surface meshes. A notable exception is [15], which, however, is not competitive in terms of surface accuracy.

In this work, we close this gap by augmenting two state-of-the-art cortical surface reconstruction (CSR) methods [1,20] with two different parcellation approaches in an end-to-end trainable manner. Namely, we extend the CSR networks with a graph classification network and, as an alternative, we propagate template parcellation labels through the CSR network via a novel class-based reconstruction loss. Both approaches are illustrated in Fig. 1. We demonstrate that both approaches yield highly accurate cortex parcellations on top of state-of-the-art boundary surfaces.

2 Related Work

In the following, we will briefly review previous work related to corical surface reconstruction and cortex parcellation. While we focus on joint reconstruction and parcellation, the majority of existing methods solves only one of these two tasks at a time, i.e., cortex parcellation *or* cortical surface reconstruction.

Convolutional neural networks (CNNs) remain a popular choice for medical image segmentation and they have been applied successfully to the task of cortex parcellation. For example, FastSurfer [16] replaces FreeSurfer's voxel-based stream by a multi-view 2D CNN. Similar approaches [3,17] have been proposed based on 3D patch-based networks. However, the computation of cortical biomarkers based on fully-convolutional segmentations is ultimately restricted by the image resolution of the input MRI scans and the combination with the FreeSurfer surface stream is not efficient in terms of inference time.

Deep learning-based parcellation methods operating on given surface meshes (typically pre-computed with FreeSurfer) have also been presented in the past. For example, the authors of [5] investigate different network architectures for

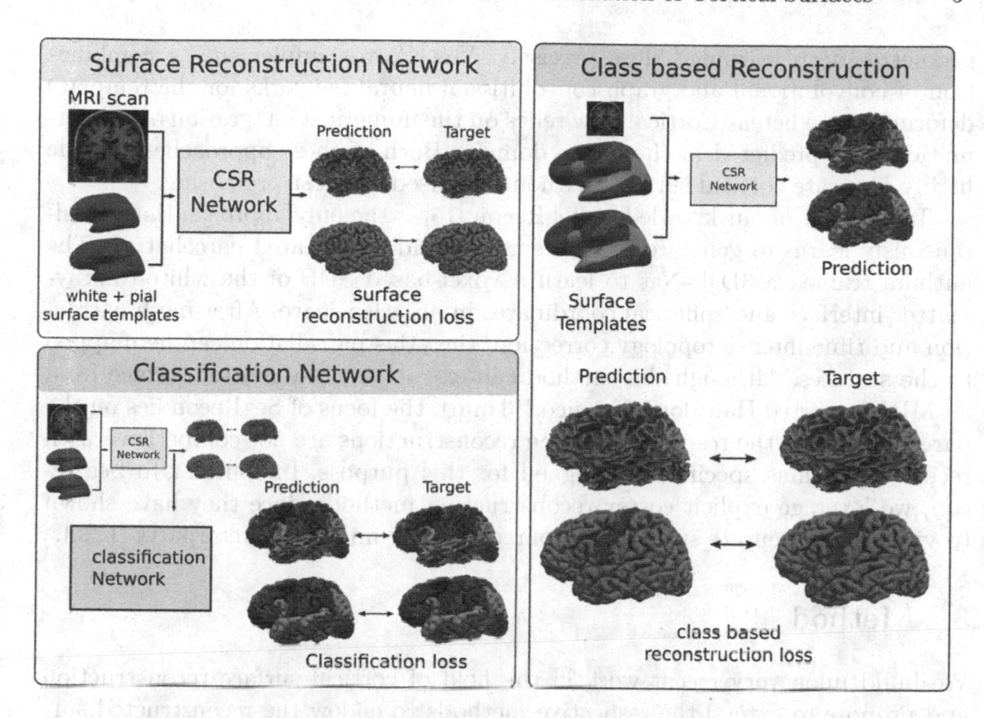

Fig. 1. Overview of surface reconstruction networks with our extensions for learning cortex parcellation. Bottom left: A classification network is added after the deformation network and trained with a classification loss on the vertex-wise class predictions. Right: The deformation network takes surface templates with parcellation labels (from a population atlas) as input and the reconstruction loss is computed separately for each class.

the segmentation of two brain areas. They found that graph convolution-based approaches are more suited compared to multi-layer perceptrons (MLPs). Similarly, the method presented in [8] parcellates the whole cortex using graph attention networks. In contrast, the authors of [14] utilize spherical graph convolutions, which they find to be more effective than graph convolutions in the Euclidean domain. All of these vertex classifiers consider the surface mesh as given.

To avoid the lengthy runtime of FreeSurfer for surface generation, deep learning-based surface reconstruction approaches focus on the fast and accurate generation of cortical surfaces from MRI. These approaches can be grouped into implicit methods [4], which learn signed distance functions (SDFs) to the white-to-gray-matter and gray-matter-to-pial interfaces, and explicit methods [1,20], which directly predict a mesh representation of the surfaces. The disadvantage of implicit surface representations is the need for intricate mesh extraction, e.g., with marching cubes [21], and topology correction. This kind of post-processing is time-consuming and can introduce anatomical errors [10]. In contrast, Vox2Cortex [1] and CorticalFlow [20] deform a template mesh based on geometric deep learning. More precisely, Vox2Cortex implements a combination of convolutional and graph-convolutional neural networks for the template deformation, whereas CorticalFlow relies on the numerical integration of a deformation field predicted in the image domain. Both of these approaches provide highly accurate cortical surfaces without the need for post-processing.

To the best of our knowledge, SegRecon [15] is the only approach that simultaneously learns to generate cortical surfaces and a dedicated parcellation. The authors trained a 3D U-Net to learn a voxel-based SDF of the white-to-gray-matter interface and spherical coordinates in an atlas space. After mesh extraction and time-intense topology correction, the atlas parcellation can be mapped to the surfaces. Although this method can extract a white matter surface from an MRI (reported Hausdorff distance 1.3 mm), the focus of SegRecon lies on the parcellation and the respective surface reconstructions are not competitive with recent algorithms specifically designed for this purpose. In contrast to SegRecon, we leverage explicit cortex reconstruction methods since they have shown to yield more accurate surfaces compared to their implicit counterparts [1,20].

3 Method

We build upon very recent work in the field of cortical surface reconstruction and propose to extend the respective methods to endow the reconstructed surfaces with a jointly learned parcellation. In particular, we base our work on Vox2Cortex [1] and CorticalFlow [20], two mesh-deformation methods that have shown state-of-the-art results for the extraction of cortical surfaces. Both of these methods take a 3D MRI scan and a mesh template as input and compute four cortical surfaces simultaneously, the white-matter and the pial surfaces of each hemisphere.

3.1 Surface Reconstruction Methods

Vox2Cortex: Inspired by previous related methods [19,30,31], Vox2Cortex (V2C) [1] consists of two neural sub-networks, a CNN that operates on voxels and a GNN responsible for mesh deformation. Both sub-networks are connected via feature-sampling modules that map features extracted by the CNN to vertex locations of the meshes. To avoid self-intersections in the final meshes, which is a common problem in explicit surface reconstruction methods, Vox2Cortex relies on multiple regularization terms in the loss function. The deformation of the template mesh is done in a sequential manner, i.e., multiple subsequent deformation steps that build upon each other lead to the final mesh prediction.

CorticalFlow: In contrast to Vox2Cortex, which predicts the mesh-deformation field on a per-vertex basis, CorticalFlow (CF) [20] relies on a deformation field in image space. To map it onto the mesh vertices, CorticalFlow interpolates the deformation field at the respective locations. Similarly to Vox2Cortex, the deformation is done step-by-step and each sub-deformation is predicted by a

3D UNet. To avoid self-intersections, the authors propose an Euler integration scheme of the flow fields. The intuition behind using a numerical integration is that by choosing a sufficiently small step size h, the mesh deformation is guaranteed to be diffeomorphic and, thus, intersection-free. However, this guarantee does not hold in practice due to the discretization of the surfaces [20]. In our experiments, we apply only a single integration step to reduce training time and memory consumption (also at training time).

3.2 Surface Parcellation

For the parcellation of the human cortex, there exist multiple atlases based on, e.g., structural or functional properties according to which different brain regions can be distinguished. Commonly used atlases are the Desikan-Killiany-Tourville (DKT) [6,18] or Destrieux [7] atlas, which are both available in FreeSurfer. We use FreeSurfer surfaces as pseudo-ground-truth meshes with parcellation labels from the DKT atlas and smoothed versions of the FreeSurfer fsaverage template as mesh input.

Classification Network: Previous work [5,8] has shown that GNN-based classification networks can provide accurate cortex parcellations. Therefore, we extend the CSR networks with a classification branch consisting of three residual GNN blocks, each with three GNN layers. We hand the predicted mesh with vertex features (extracted by the Vox2Cortex GNN) or just vertex coordinates (from CorticalFlow) as input to the GNN classifier. As output, we obtain a vector of class probabilities for each vertex. After a softmax layer, we compute a cross-entropy loss between the predicted classes and ground-truth classes of the closest points in the target mesh. In combination with Vox2Cortex, we integrate the classification network after the last mesh-deformation step and train the CSR and classification networks end-to-end. In combination with CorticalFlow, we also add the classification layer after the last deformation and freeze the parameters of the U-Nets of the previous steps. In our experiments, we found that adding the classification network in each of the iterative optimization steps leads to training instability, hence we only add it in the last iteration.

Class-Based Reconstruction: As both Vox2Cortex and CorticalFlow are template-deformation approaches, we propose to propagate the atlas labels of the DKT atlas through the deformation process. More precisely, we enforce the respective regions from the template to fit the labeled regions of the FreeSurfer meshes by using a modified *class-based* reconstruction loss. This loss function is agnostic to the concrete implementation of the reconstruction loss, e.g., it can be given by a Chamfer distance as in CorticalFlow or a combination of point-weighted Chamfer and normal distance as in Vox2Cortex. Let $\mathcal{P}^{\mathrm{p}}$ and $> \mathcal{P}$ be predicted and ground-truth point sets sampled from the meshes $\mathcal{M}^{\mathrm{p}}$ and $> \mathcal{M}$, potentially associated with normals. Further, let $\mathcal{L}_{\mathrm{rec}}(\mathcal{P}^{\mathrm{p}}_c, > \mathcal{P}_c)$ be any reconstruction loss between the point clouds of a certain parcellation class $c \in C$. Then, we compute the class-based reconstruction loss as

$$\mathcal{L}_{\text{rec},class}(\mathcal{P}^{\text{p}}, > \mathcal{P}) = \frac{1}{|C|} \sum_{c \in C} \mathcal{L}_{\text{rec}}(\mathcal{P}^{\text{p}}_c, > \mathcal{P}_c). \tag{1}$$

Intuitively, the predicted points of a certain atlas class "see" only the ground-truth points of the same class. We depict this intuition also in Fig. 1B. By construction of this loss, the parcellation of the deformed template and the ground-truth parcellation are aligned. Compared to the classification network, this approach has the advantage that "islands", i.e., small wrongly classified regions, cannot occur on smooth reconstructed meshes.

3.3 Experimental Setup

Data: We train our models on 292 subjects of the publicly available OASIS-1 dataset [25] and use 44 and 80 subjects for validation and testing, respectively. Overall, 100 subjects in OASIS-1 have been diagnosed with very mild to moderate Alzheimer's disease. We based our splits on diagnosis, age, and sex in order to avoid training bias against any of these groups.
Pre-processing: We use the FreeSurfer software pipeline, version v7.2[1], as silver standard for training and evaluation of our models. More precisely, we follow the setup in [1,20] and use the orig.mgz files and white and pial surfaces generated by FreeSurfer and register the MRI scans to the MNI152 space (rigid and subsequent affine registration). Further, we subsampled the surface meshes to about 40,000 vertices per surface using quadric edge collapse decimation [11]. Images are padded to have shape 192×208×192. For Vox2Cortex experiments, we resize the images after padding to 128×144×128 voxels as done in the original paper [1]. We use min-max-normalization of intensity values to scale to $[0, 1]$.
Training: For the computation of the reconstruction losses, we sample 50,000 points from the predicted and ground-truth meshes in a differentiable manner [12]. We interpolate curvature information of a sampled point using the barycentric weights from the respective triangle vertices and assign the point class of the closest vertex to a sampled point. For training CorticalFlow, we use an iterative procedure as described by [20], i.e., freezing the UNet(s) of steps 1 to $i-1$ when training UNet i. We further use the AdamW optimizer [22] with weight decay 1e−4 and a cyclic learning rate schedule [29] for the optimization of the networks. As input to the deformation networks, we leverage the fsaverage templates in FreeSurfer and smooth them extensively using the HC Laplacian smoothing operator implemented in MeshLab [2]. We provide a list of all model parameters, which we adopt from the Vox2Cortex and CorticalFlow papers, in the supplemental material. Our implementation is based on PyTorch [26] and PyTorch3d [27] and we trained on an Nvidia Titan RTX GPU.

4 Results

In the following, we show results for both of the proposed parcellation approaches, i.e., the classification network and the class-based reconstruction.

[1] Available at https://surfer.nmr.mgh.harvard.edu/.

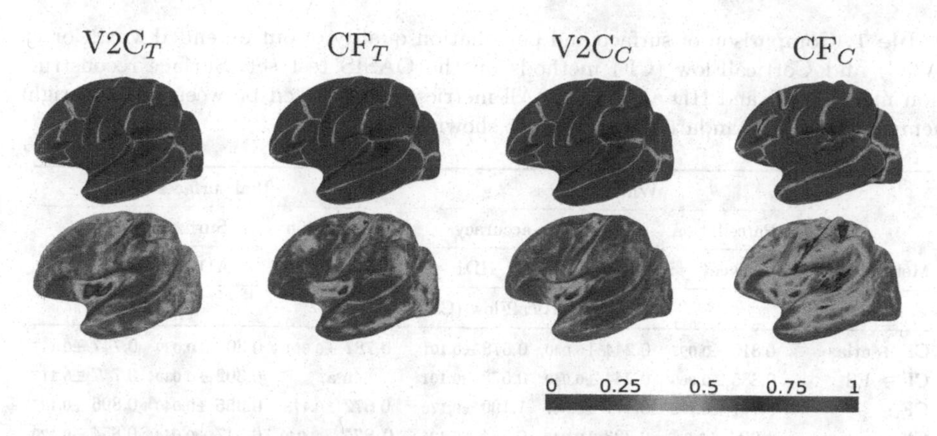

Fig. 2. Visualization of parcellation and reconstruction accuracy for our four methods, averaged over the predicted left white surfaces of the OASIS test set. Top row: parcellation error, blue (0.0) = vertex classified correctly for all subjects, red (1.0) = vertex classified incorrectly for all subjects. Errors are mostly present in parcel boundaries. Bottom row: average distance from predicted surface to ground-truth surface, in mm. (Color figure online)

To this end, we combine the proposed methods with Vox2Cortex (V2C) and CorticalFlow (CF) as described in Sect. 3. This leads to a total of four methods, which we denote as $V2C_C$, CF_C (classification network) and $V2C_T$, CF_T (class-based reconstruction via template). We compare our approaches to FastSurfer [16] and two additional baselines per reconstruction method. The latter are obtained by (1) training "vanilla" CF and V2C and mapping the atlas labels simply to the predicted surfaces (denoted as CF + atlas and V2C + atlas) and (2) using FreeSurfer's spherical atlas registration as an ad-hoc parcellation of given surfaces in a post-processing fashion (denoted as CF + FS and V2C + FS). The runtime for FastSurfer is about one hour, for the FS parcellation of CF and V2C meshes several hours, and for the proposed methods in the range of seconds. Table 1 presents the parcellation accuracy in terms of average Dice coefficient over all parcellation classes (computed on the surfaces). In addition, we compare the surface reconstruction accuracy in terms of average symmetric surface distance (AD) and 90th percentile Hausdorff distance (HD) in mm.

We observe that FastSurfer leads to highly accurate surfaces compared to the FreeSurfer silver standard, which is expected as FastSurfer makes use of the FreeSurfer surface stream to generate surfaces. The baseline CF and V2C models also provide very accurate predictions in terms of surface accuracy with a slight advantage on the side of CF (probably due to the higher image and mesh resolution at training time). However, as expected, CF and V2C do not yield an accurate surface parcellation if a population atlas is used as their input. Generating the DKT parcellation with FreeSurfer's atlas registration yields a higher Dice score than FastSurfer, which we attribute to the superiority of the mesh-based parcellation compared to a voxel-based approach. Note that the

Table 1. Comparison of surface and parcellation quality of our extended Vox2Cortex (V2C) and CorticalFlow (CF) methods on the OASIS test set. Surface reconstruction metrics AD and HD are in mm. All metrics are averaged between left and right hemispheres and standard deviations are shown.

	White surfaces			Pial surfaces		
	Parcellation	Surface accuracy		Parcellation	Surface accuracy	
Method	Dice↑	AD↓	HD↓	Dice↑	AD↓	HD↓
CorticalFlow (CF) [20]						
CF + atlas	0.810 ±0.095	0.244 ±0.040	0.578 ±0.101	0.787 ±0.091	0.302 ±0.039	0.747 ±0.117
CF + FS	0.885 ±0.069	0.244 ±0.040	0.578 ±0.101	n.a.	0.302 ±0.039	0.747 ±0.117
CF_C	0.727 ±0.178	0.471 ±0.047	1.190 ±0.170	0.672 ±0.178	0.355 ±0.040	0.896 ±0.126
CF_T	**0.904** ±0.048	0.323 ±0.048	0.784 ±0.126	**0.877** ±0.049	0.347 ±0.044	0.854 ±0.120
Vox2Cortex (V2C) [1]						
V2C + atlas	0.740 ±0.121	0.282 ±0.034	0.587 ±0.078	0.691 ±0.132	0.341 ±0.037	0.848 ±0.124
V2C + FS	0.876 ±0.076	0.282 ±0.034	0.587 ±0.078	n.a.	0.341 ±0.037	0.848 ±0.124
$V2C_C$	**0.902** ±0.050	0.303 ±0.034	0.641 ±0.082	**0.876** ±0.053	0.362 ±0.038	0.894 ±0.119
$V2C_T$	0.885 ±0.057	0.372 ±0.051	0.823 ±0.108	0.858 ±0.058	0.429 ±0.052	1.066 ±0.182
FastSurfer [16]	0.862 ±0.084	0.138 ±0.057	0.331 ±0.172	0.839 ±0.081	0.240 ±0.065	0.557 ±0.179

FreeSurfer spherical registration only works on white matter surfaces and, thus, is not applicable for the parcellation of pial surfaces.

Regarding the surface quality, we observe that solving the additional task of cortex parcellation causes a slight loss of surface accuracy in all models. This effect is most severe in the CF_C and $V2C_T$ models. As the mesh-deformation network in V2C already requires several regularization losses, we suspect that the restrictive class-based reconstruction loss might interfere with the regularizers. In terms of parcellation accuracy, we observe best results for CF_T and $V2C_C$ models with an average Dice score greater than 0.9 for white surfaces and 0.87 for pial surfaces over all parcels. The classification GNN in $V2C_C$ takes the vertex features of the previous GNNs as input. Consequently, it can make use of vertex-wise information, which is not available in CF_C (in this case, the classification network only gets vertex locations as input). As expected, CF_C yields a lower parcellation accuracy compared to $V2C_C$. Therefore, we conclude that a combination of CF with a GNN classification network is not an optimal choice.

We visualize the parcellation and surface reconstruction accuracy of the left white surfaces in Fig. 2 and observe that, averaged over the test set, classification errors occur almost exclusively at parcel boundaries. Visualizations of pial surfaces are shown in the supplement. Overall, we conclude that the GNN classifier is better suited for V2C than for CF, as the previous graph convolutions provide more meaningful vertex input features. In contrast, the class-based reconstruction loss leads to better results in CF.

5 Conclusion

In this work, we introduced two effective extensions to brain reconstruction networks for joint cortex parcellation: one based on a graph classifier and one based on a novel and generic region-based reconstruction loss. Both methods are particularly suited to augment mesh-deformation networks, which provide highly accurate surface meshes, with the ability to parcellate the surfaces into associated regions. The extremely short runtime of the presented algorithms, which lies in the range of seconds at inference time, together with the high parcellation accuracy paves the way for a more fine-grained analysis of brain diseases in large-cohort studies and the integration in clinical practice.

Acknowledgments. This research was partially supported by the Bavarian State Ministry of Science and the Arts and coordinated by the bidt, and the BMBF (DeepMentia, 031L0200A). We gratefully acknowledge the computational resources provided by the Leibniz Supercomputing Centre (www.lrz.de).

References

1. Bongratz, F., Rickmann, A.M., Pölsterl, S., Wachinger, C.: Vox2cortex: fast explicit reconstruction of cortical surfaces from 3D MRI scans with geometric deep neural networks. In: CVPR (2022)
2. Cignoni, P., Callieri, M., Corsini, M., Dellepiane, M., Ganovelli, F., Ranzuglia, G.: MeshLab: an open-source mesh processing tool. In: Eurographics Italian Chapter Conference. The Eurographics Association (2008)
3. Coupé, P., et al.: Assemblynet: a large ensemble of CNNs for 3Dd whole brain MRI segmentation. NeuroImage **219**, 117026 (2020)
4. Cruz, R.S., Lebrat, L., Bourgeat, P., Fookes, C., Fripp, J., Salvado, O.: Deepcsr: a 3D deep learning approach for cortical surface reconstruction. In: WACV, pp. 806–815 (2021)
5. Cucurull, G., et al.: Convolutional neural networks for mesh-based parcellation of the cerebral cortex (2018)
6. Desikan, R.S., et al.: An automated labeling system for subdividing the human cerebral cortex on MRI scans into gyral based regions of interest. NeuroImage **31**(3), 968–980 (2006)
7. Destrieux, C., Fischl, B., Dale, A., Halgren, E.: Automatic parcellation of human cortical gyri and sulci using standard anatomical nomenclature. NeuroImage **53**(1), 1–15 (2010)
8. Eschenburg, K.M., Grabowski, T.J., Haynor, D.R.: Learning cortical parcellations using graph neural networks. Front. Neurosci. **15** (2021)
9. van Essen, D.C., Glasser, M.F., Dierker, D.L., Harwell, J.W., Coalson, T.S.: Parcellations and hemispheric asymmetries of human cerebral cortex analyzed on surface-based atlases. Cerebral Cortex **22**(10), 2241–2262 (2012)
10. Fischl, B.: Freesurfer. Neuroimage **62**(2), 774–781 (2012)
11. Garland, M., Heckbert, P.S.: Surface simplification using quadric error metrics. In: Proceedings of the 24th Annual Conference on Computer Graphics and Interactive Techniques (SIGGRAPH 1997), USA, pp. 209–216 (1997)
12. Gkioxari, G., Johnson, J., Malik, J.: Mesh r-cnn. In: ICCV, pp. 9784–9794 (2019)

13. Glasser, M.F., et al.: A multi-modal parcellation of human cerebral cortex. Nature **536**(7615), 171–178 (2016)
14. Gopinath, K., Desrosiers, C., Lombaert, H.: Graph convolutions on spectral embeddings for cortical surface parcellation. Med. Image Anal. **54**, 297–305 (2019)
15. Gopinath, K., Desrosiers, C., Lombaert, H.: SegRecon: learning joint brain surface reconstruction and segmentation from images. In: de Bruijne, M., et al. (eds.) MICCAI 2021. LNCS, vol. 12907, pp. 650–659. Springer, Cham (2021). https://doi.org/10.1007/978-3-030-87234-2_61
16. Henschel, L., Conjeti, S., Estrada, S., Diers, K., Fischl, B., Reuter, M.: Fastsurfer - a fast and accurate deep learning based neuroimaging pipeline. NeuroImage **219**, 117012 (2020)
17. Huo, Y., et al.: 3D whole brain segmentation using spatially localized atlas network tiles. NeuroImage **194**, 105–119 (2019)
18. Klein, A., Tourville, J.: 101 labeled brain images and a consistent human cortical labeling protocol. Front. Neurosci. **6** (2012)
19. Kong, F., Shadden, S.C.: Whole heart mesh generation for image-based computational simulations by learning free-from deformations (2021)
20. Lebrat, L., et al.: Corticalflow: a diffeomorphic mesh transformer network for cortical surface reconstruction. Adv. Neural Inf. Process. Syst. **34** (2021)
21. Lewiner, T., Lopes, H., Vieira, A.W., Tavares, G.: Efficient implementation of marching cubes' cases with topological guarantees. J. Graph. Tools **8**(2), 1–15 (2003)
22. Loshchilov, I., Hutter, F.: Decoupled weight decay regularization. In: International Conference on Learning Representations (2019)
23. Ma, Q., Robinson, E.C., Kainz, B., Rueckert, D., Alansary, A.: PialNN: a fast deep learning framework for cortical pial surface reconstruction. In: Abdulkadir, A., et al. (eds.) MLCN 2021. LNCS, vol. 13001, pp. 73–81. Springer, Cham (2021). https://doi.org/10.1007/978-3-030-87586-2_8
24. Mai, J.K., Paxinos, G.: The Human Nervous System. Academic Press (2011)
25. Marcus, D.S., Wang, T.H., Parker, J., Csernansky, J.G., Morris, J.C., Buckner, R.L.: Open access series of imaging studies (OASIS): cross-sectional MRI data in young, middle aged, nondemented, and demented older adults. J. Cognit. Neurosci. **19**(9), 1498–1507 (2007)
26. Paszke, A., et al.: Pytorch: an imperative style, high-performance deep learning library. Adv. Neural Inf. Process. Syst. **32**, 8024–8035 (2019)
27. Ravi, N., et al.: Accelerating 3d deep learning with pytorch3d. arXiv:2007.08501 (2020)
28. Roe, J.M., et al.: Asymmetric thinning of the cerebral cortex across the adult lifespan is accelerated in Alzheimer's disease. Nat. Commun. **12**(1) (2021)
29. Smith, L.N.: Cyclical learning rates for training neural networks. In: WACV, pp. 464–472 (2017)
30. Wang, N., Zhang, Y., Li, Z., Fu, Y., Liu, W., Jiang, Y.-G.: Pixel2Mesh: generating 3D mesh models from single RGB images. In: Ferrari, V., Hebert, M., Sminchisescu, C., Weiss, Y. (eds.) ECCV 2018. LNCS, vol. 11215, pp. 55–71. Springer, Cham (2018). https://doi.org/10.1007/978-3-030-01252-6_4
31. Wickramasinghe, U., Remelli, E., Knott, G., Fua, P.: Voxel2Mesh: 3D mesh model generation from volumetric data. In: Martel, A.L., et al. (eds.) MICCAI 2020. LNCS, vol. 12264, pp. 299–308. Springer, Cham (2020). https://doi.org/10.1007/978-3-030-59719-1_30

A Study of Demographic Bias in CNN-Based Brain MR Segmentation

Stefanos Ioannou[1(✉)], Hana Chockler[1,3], Alexander Hammers[2], Andrew P. King[2], and for the Alzheimer's Disease Neuroimaging Initiative

[1] Department of Informatics, King's College London, London, UK
stefanos.ioannou@kcl.ac.uk

[2] School of Biomedical Engineering and Imaging Sciences, King's College London, London, UK

[3] causaLens Ltd., London, UK

Abstract. Convolutional neural networks (CNNs) are increasingly being used to automate the segmentation of brain structures in magnetic resonance (MR) images for research studies. In other applications, CNN models have been shown to exhibit bias against certain demographic groups when they are under-represented in the training sets. In this work, we investigate whether CNN models for brain MR segmentation have the potential to contain sex or race bias when trained with imbalanced training sets. We train multiple instances of the FastSurfer-CNN model using different levels of sex imbalance in white subjects. We evaluate the performance of these models separately for white male and white female test sets to assess sex bias, and furthermore evaluate them on black male and black female test sets to assess potential racial bias. We find significant sex and race bias effects in segmentation model performance. The biases have a strong spatial component, with some brain regions exhibiting much stronger bias than others. Overall, our results suggest that race bias is more significant than sex bias. Our study demonstrates the importance of considering race and sex balance when forming training sets for CNN-based brain MR segmentation, to avoid maintaining or even exacerbating existing health inequalities through biased research study findings.

Keywords: Brain · MR · Deep learning · Bias · Fairness

Data used in preparation of this article were obtained from the Alzheimer's Disease Neuroimaging Initiative (ADNI) database (adni.loni.usc.edu). As such, the investigators within the ADNI contributed to the design and implementation of ADNI and/or provided data but did not participate in analysis or writing of this report. A complete listing of ADNI investigators can be found at: http://adni.loni.usc.edu/wp-content/uploads/how_to_apply/ADNI_Acknowledgement_List.pdf.

Supplementary Information The online version contains supplementary material available at https://doi.org/10.1007/978-3-031-17899-3_2.

A. Abdulkadir et al. (Eds.): MLCN 2022, LNCS 13596, pp. 13–22, 2022.
https://doi.org/10.1007/978-3-031-17899-3_2

1 Introduction

The study of bias and fairness in artificial intelligence (AI) has already attracted significant interest in the research community, with the majority of studies considering fairness in classification tasks in computer vision [14]. There are many causes of bias in AI, but one of the most common is the combination of imbalance and distributional shifts in the training data between protected groups[1]. For example, [3] found bias in commercial gender classification models caused by under-representation of darker-skinned people in the training set. Recently, a small number of studies have investigated bias in AI models for medical imaging applications. For example, [2,11] found significant under-performance on chest X-ray diagnostic models when evaluated on protected groups such as women that were under-represented in the training data. [11] concluded that training set diversity and gender balance is essential for minimising bias in AI-based diagnostic decisions. Similarly, [1] found bias in AI models for skin lesion classification and proposed a debiasing technique based on an adversarial training method.

Whilst in computer vision classification tasks are commonplace, in medicine image segmentation plays a crucial role in many clinical workflows and research studies. For example, segmentation can be used to quantify cardiac function [18] or to understand brain anatomy and development [6]. AI techniques are increasingly being used to automate the process of medical image segmentation [8]. For example, in the brain techniques based upon convolutional neural networks (CNNs) have been proposed for automatically segmenting magnetic resonance (MR) images [5,6], outperforming the previous state of the art. However, the only study to date to have investigated bias in segmentation tasks has been [15,16], which found significant racial bias in the performance of a CNN model for cardiac MR segmentation, caused by racial imbalance in the training data.

The structural anatomy of the brain is known to vary between different demographic groups, such as sex [4] and race [7]. Given that a known cause of bias in AI is the presence of such distributional shifts, and the increasing use of AI-based segmentation tools in brain imaging, it is perhaps surprising that no study to date has investigated the potential for bias in AI-based segmentation of the brain. In this paper we perform such a study. We first systematically vary levels of sex imbalance in a training set of white subjects to train multiple instances of the FastSurferCNN AI segmentation model [6]. We evaluate the performance of these models separately using test sets comprised of white male and white female subjects to assess potential sex bias. Subsequently, we assess potential race bias by evaluating the performance of the same models on black male and black female subjects.

---

[1] A *protected group* is a set of samples which all share the same value of the *protected attribute*. A protected attribute is one where fairness needs to be guaranteed, e.g. race and sex.

2 Materials and Methods

2.1 Data

To evaluate potential sex and race bias, we used MR images from the Alzheimer's Disease Neuroimaging Initiative (ADNI)[2] database. We used a total of 715 subjects consisting of males and females of white or black race (according to the sex and race information stored in the ADNI database).

We used ground truth segmentations produced by the Multi-Atlas Label Propagation with Expectation-Maximisation based refinement (MALPEM) algorithm[3] [12,13]. We chose MALPEM ground truth segmentations due to the lack of manual ground truths with sufficient race/sex representation for our experiments, and the fact that MALPEM performed accurately and reliably in an extensive comparison on clinical data [9]. MALPEM segments the brain into 138 anatomical regions. See Table S2 in the Supplementary Materials for a description of the MALPEM regions.

2.2 Model and Training

For our experiments, we used the FastSurferCNN model, which is part of the FastSurfer pipeline introduced in [6]. The authors of the model used training data that labelled the brain following the Desikan-Killiany-Tourville (DKT) atlas, and reduced the number of anatomical regions to 78, by lateralising or combining cortical regions that are in contact with each other across hemispheres. We follow a similar approach, by lateralising, removing or combining cortical regions in the MALPEM segmentations to retain the same number of anatomical structures (i.e. 78), thus enabling us to use the FastSurferCNN model without modifying its architecture. See Table S2 in the Supplementary Materials for details of how we reduced the number of anatomical regions in the MALPEM segmentations to be consistent with those expected by the FastSurferCNN model.

Models were trained using the coronal slices of the 3D MR data. The training procedure followed was identical to that of [6]. All models were trained for 30 epochs, with a learning rate of 0.01 decreased by a factor of 0.3 every 5 epochs, a batch size of 16 and using the Adam optimiser [10]. Random translation was used to augment the training set as in [6]. The loss function combines logistic loss and Dice Loss [17]. A validation set of 10–20 subjects was used to monitor the training procedure. For each comparison we repeated the training twice, and report the average performance over the two runs.

The training procedure resulted in average Dice Similarity Coefficients (AVG DSC) similar to that of [6], therefore we assume models were trained to their full capacity.

---

[2] www.adni-info.org.

[3] We used the segmentations available at https://doi.gin.g-node.org/10.12751/g-node.aa605a/.

3 Experiments

We now describe the experiments that we performed to investigate possible sex and race bias in the use of the FasterSurferCNN model for brain MR segmentation. We describe experiments to assess sex and race bias separately below.

3.1 Sex Bias

When analysing sex bias, to remove the potential confounding factor of race we used only white subjects since they had the largest number of subjects in the ADNI database.

Table 1 lists the datasets used for training/evaluation to investigate sex bias. Models were trained using training sets with different proportions of white male and white female subjects. All of the models were evaluated on the same 185 white male and 185 white female subjects. For all datasets the scanner manufacturer (Siemens, GE Medical Systems, and Philips), subject age, field strength (3.0T and 1.5T), and diagnosis (Dementia, Mild Cognitive Impairment, and Cognitively Normal) were controlled for. All images were acquired using the MP-RAGE sequence, which showed the highest performance in [6].

Table 1. Training and test sets used to assess sex bias. The number (and proportion) of white female and white male subjects and the mean and standard deviations (SD) of the subjects' ages in each dataset.

Usage	Female, n (%)	Male, n (%)	Age ± SD	Description
Training	140 (100)	0 (0)	74.29 ± 5.52	100% female
	35 (25)	105 (75)	75.6 ± 4.82	75% male, 25% female
	70 (50)	70 (50)	75.13 ± 5.34	50% male, 50% female
	105 (75)	35 (25)	74.64 ± 5.64	25% male, 75% female
	0 (0)	140 (100)	75.68 ± 4.84	100% male
Testing	185 (100)	0 (0)	74.44 ± 5.16	Female test set
	0	185 (100)	75.55 ± 5.55	Male test set

3.2 Racial Bias

To investigate racial bias, the same models trained with white subjects in the sex bias experiment (see Table 1) were evaluated on test sets broken down by both sex and race, utilising an extra set of male/female black race subjects. We used four test sets in total in this experiment: white female, white male, black female and black male. All test sets consisted of 36 subjects each and were controlled for age, scanner manufacturer, field strength and diagnosis as in the sex bias experiment.

3.3 Evaluation and Statistical Analysis

To evaluate the performance of the models, we computed the DSC on a per-region basis as well as the generalised DSC (GDSC) [19] (excluding the background class) to quantify overall performance. All metrics were computed for each test set individually as detailed in Sects. 3.1 and 3.2. Statistical tests were performed using Wilcoxon signed rank tests (for paired comparisons) and Mann-Whitney U tests (for unpaired comparisons), both using ($p \leq 0.01$) significance between DSC values for different models.[4]

4 Results

4.1 Sex Bias

The AVG GDSC results (see Fig. S1 in the Supplementary Materials) showed that the models exhibit some signs of potential bias for both white males and white females. However, most of the comparisons showed no statistical significance when considering the GDSC overall. Therefore, we more closely analysed individual brain regions that did show statistically significant differences in performance in terms of per-region AVG DSC.

Figure 1 shows the performance of the models for selected regions exhibiting the highest bias when evaluated on white female subjects. As can be seen, AVG DSC for these regions decreases as the proportion of female subjects in the training set decreases. Over all regions, when evaluating on white females the model trained using 100% male subjects had significantly lower AVG DSC compared to the model trained with 100% females in 53 of the 78 regions ($p \leq 0.01$). The Both-PCgG-posterior-cingulate-gyrus region exhibits the highest decrease in AVG-DSC of 0.0395. Similarly, when evaluating on white males (Fig. 2), the AVG DSC for regions showing the highest bias decreases with the proportion of males in the training set. This time, 36 of the 78 regions had significantly worse performance for the 100% female trained model compared to the 100% male trained model ($p \leq 0.01$). The Both-OCP-occipital-pole region exhibits the highest decrease in AVG-DSC of 0.0406. For both the white male and white female test sets, no region showed statistical significance in the opposite direction. Interestingly, we note that for the majority of regions, all models perform better on males than on females, even when the model is trained only on female subjects. For instance, the model trained with a sex-balanced dataset shows a significant decrease ($p \leq 0.01$ using the Mann-Whitney U test) in 3 regions when evaluated on white female subjects compared to white male subjects, with no significant difference in the opposite direction. The Both-Inf-Lat-Vent region shows the highest decrease in AVG-DSC of 0.0475.

[4] When assessing differences for multiple regions we did not apply correction for multiple tests because our aim was to be sensitive to possible bias rather than minimise Type I errors.

4.2 Racial Bias

For assessing racial bias, the AVG GDSC results (see Fig. S2 in the Supplementary Materials) show that the models again exhibit some signs of bias but most

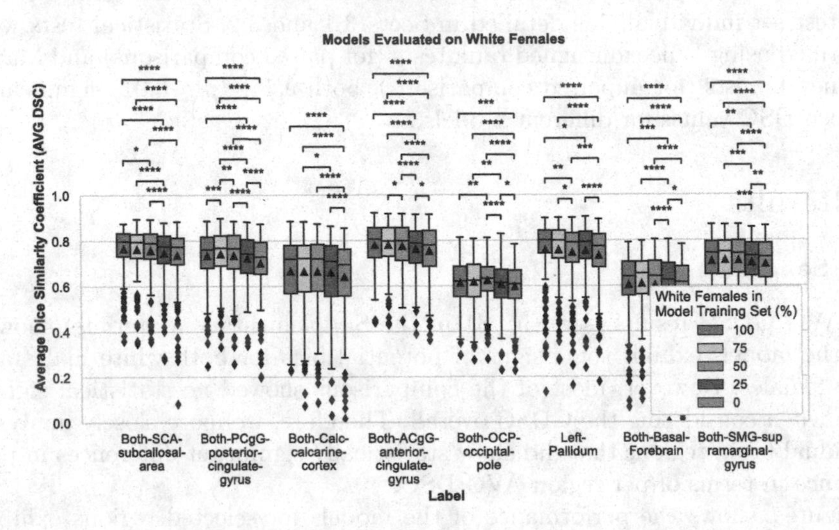

Fig. 1. Regions showing the highest bias according to models' DSC performance on white females. Significance using Wilcoxon signed ranked test is denoted by **** ($P \leq 1.00e-04$), *** ($1.00e-04 < P \leq 1.00e-03$), ** ($1.00e-03 < P \leq 1.00e-02$), and * ($1.00e-02 < P \leq 5.00e-02$).

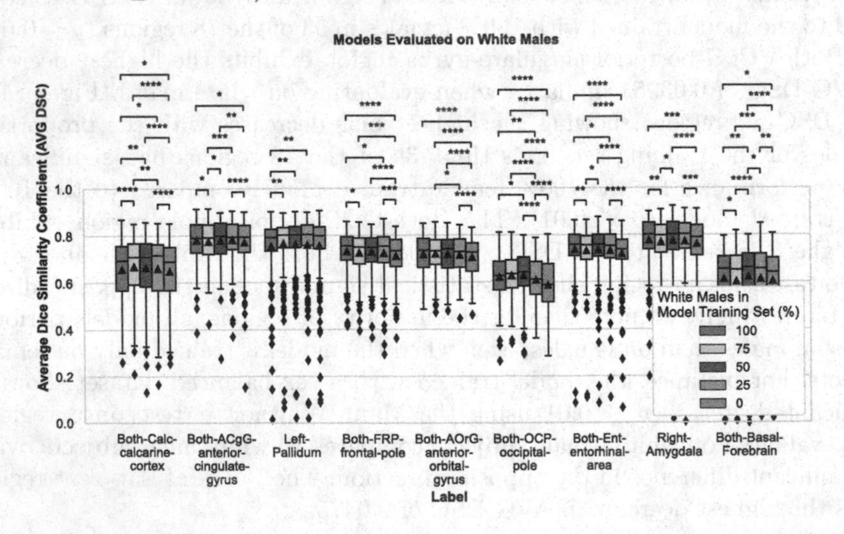

Fig. 2. Regions showing the highest bias according to models' DSC performance on white males. Significance using Wilcoxon signed ranked test is denoted by **** ($P \leq 1.00e-04$), *** ($1.00e-04 < P \leq 1.00e-03$), ** ($1.00e-03 < P \leq 1.00e-02$), and * ($1.00e-02 < P \leq 5.00e-02$).

comparisons are not statistically significant. However, it can be observed that statistically significant differences are found when using a black female test set.

As in the sex bias experiments, we more closely analysed individual regions that showed the highest bias in AVG DSC. Figure 3 summarises these results for white and black female test sets. We can see a significant drop in performance when the (white-trained) models are evaluated on black female subjects. This difference in performance becomes more pronounced as the proportion of female subjects decreases, indicating a possible interaction between sex and race bias. When comparing test performance on black female subjects compared to white female subjects, the model trained only on (white) female subjects exhibits the highest decrease in AVG-DSC of 0.0779 in the Both-PoG-postcentral-gyrus region. The model trained only on (white) males shows the highest decrease in AVG-DSC of 0.0868 in the Both-SMC-supplementary-motor-cortex region. No region showed statistically significantly higher AVG DSC in black subjects compared to white, for either sex.

We also performed a comparison (see Fig. 4) between the effects of sex and race bias and found that the race bias effect was more significant than the sex bias effect.

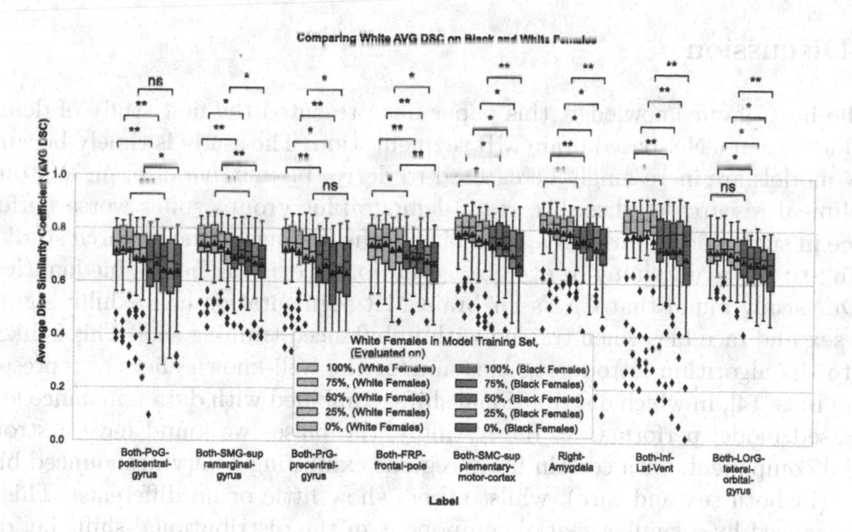

Fig. 3. Comparing AVG DSC performance of models trained with white subjects, evaluated on white and black female subjects. Mann-Whitney U test is denoted by **** ($P \leq 1.00e{-}04$), *** ($1.00e{-}04 < P \leq 1.00e{-}03$), ** ($1.00e{-}03 < P \leq 1.00e{-}02$), * ($1.00e{-}02 < P \leq 5.00e{-}02$), and not significant (ns) ($P > 5.00e{-}02$).

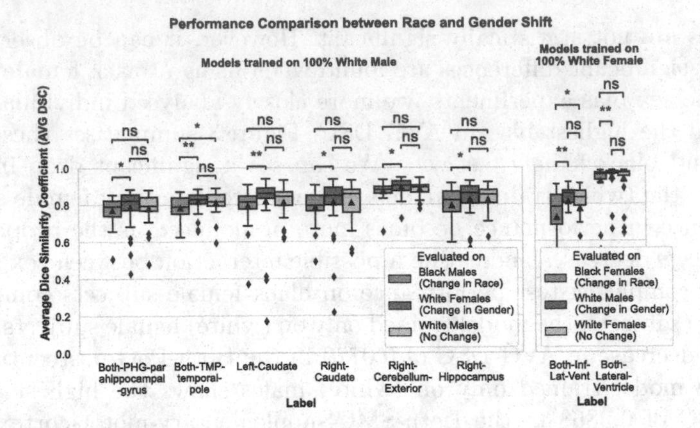

Fig. 4. Comparison of gender and race bias. The plots include all labels with a statistically significant change in performance by sex or race, of models trained on white males (left) and white females (right). Significance using Mann-Whitney U test is denoted by ** ($1.00e{-}03 < P \leq 1.00e{-}02$), * ($1.00e{-}02 < P \leq 5.00e{-}02$), and not significant (ns) ($P > 5.00e{-}02$).

5 Discussion

To the best of our knowledge, this paper has presented the first study of demographic bias in CNN-based brain MR segmentation. The study is timely because CNN models are increasingly being used to derive biomarkers of brain anatomy for clinical research studies. If certain demographic groups suffer worse performance in such models this will lead to bias in the findings of the research studies, leading to the maintaining or even exacerbation of existing health inequalities.

Our study found that CNNs for brain MR segmentation can exhibit significant sex and race bias when trained with imbalanced training sets. This is likely due to the algorithm introducing bias and/or the well-known effect of representation bias [14], in which distributional shifts combined with data imbalance lead to biased model performance. Interestingly, the biases we found have a strong spatial component, with certain brain regions exhibiting a very pronounced bias effect (in both sex and race), whilst others show little or no difference. This is likely caused by a similar spatial component in the distributional shift, i.e. differences in brain anatomy between sexes and races are likely to be localised to certain regions. We found that sex bias in performance still exists even when the model's training set is sex balanced. This is likely due to algorithmic rather than representation bias. Overall, we found that the effect of race bias was stronger than that of sex bias. Furthermore, the bias effect was much more pronounced in black females than black males.

We also observed that there was not always a monotonically increasing/decreasing trend in the performance of the models as we changed the level of imbalance. In particular, it was often the case that models trained with a (small) proportion of a different protected group improved performance for the majority group. We speculate that including data from a different protected group(s) can increase diversity in the training set, hence improving the generalisation ability of the model and leading to better performance for all protected groups.

We believe that our findings are important for the future use of CNNs in clinical research based on neuroimaging. However, we acknowledge a number of limitations. First, the number of subjects we could employ in the study was necessarily limited. The majority of subjects in the ADNI database with available MALPEM ground truth segmentations are Caucasian (i.e. white) and so we were limited in the numbers of non-white subjects we could make use of. This prevented us from performing a systematic study of the impact of race imbalance, similar to the way in which we trained multiple models with varying sex imbalance in the sex bias experiment. Another limitation is that we could not employ manual ground truth segmentations in our study. Again, this was because of the lack of large numbers of manual ground truth segmentations that are publicly available, particularly for non-white races. Although the MALPEM segmentations used in this study have been quality-inspected, we cannot assume that they are unbiased according to race and sex, or free of other systematic and random errors. Even so, we believe that the disparities in performance observed in our study can be regarded as model-induced bias, perhaps in addition to that which might be present in the ground-truth segmentations.

Another limitation is that, in this work, we only used the coronal orientation of the 3-D MR images for training and evaluation of the models. Training a model using all orientations will be the subject of future work. Future work will also focus on more extensive and detailed studies of demographic bias in brain MR segmentation, as well as investigation of techniques for bias mitigation [14].

References

1. Abbasi-Sureshjani, S., Raumanns, R., Michels, B.E.J., Schouten, G., Cheplygina, V.: Risk of training diagnostic algorithms on data with demographic bias. In: Cardoso, J., et al. (eds.) IMIMIC/MIL3ID/LABELS -2020. LNCS, vol. 12446, pp. 183–192. Springer, Cham (2020). https://doi.org/10.1007/978-3-030-61166-8_20
2. Banerjee, I., et al.: Reading Race: AI Recognises Patient's Racial Identity in Medical Images (2021)
3. Buolamwini, J.: Gender shades: intersectional accuracy disparities in commercial gender classification*. Proc. Mach. Learn. Res. **81**, 1–15 (2018)
4. Cosgrove, K.P., Mazure, C.M., Staley, J.K.: Evolving knowledge of sex differences in brain structure, function, and chemistry. Biol. Psychiat. **62**(8), 847–855 (2007). https://doi.org/10.1016/j.biopsych.2007.03.001
5. Coupé, P., et al.: AssemblyNet: a large ensemble of CNNs for 3D whole brain MRI segmentation. NeuroImage **219**, 117026 (2020). https://doi.org/10.1016/J.NEUROIMAGE.2020.117026
6. Henschel, L., Conjeti, S., Estrada, S., Diers, K., Fischl, B., Reuter, M.: FastSurfer - a fast and accurate deep learning based neuroimaging pipeline. NeuroImage **219**, 117012 (2020). https://doi.org/10.1016/j.neuroimage.2020.117012
7. Isamah, N., et al.: Variability in frontotemporal brain structure: the importance of recruitment of African Americans in neuroscience research. PLoS ONE **5**(10), e13642 (2010). https://doi.org/10.1371/journal.pone.0013642

8. Isensee, F., Jaeger, P.F., Kohl, S.A., Petersen, J., Maier-Hein, K.H.: nnU-Net: a self-configuring method for deep learning-based biomedical image segmentation. Nat. Methods **18**(2), 203–211 (2021). https://doi.org/10.1038/s41592-020-01008-z
9. Johnson, E.B., et al.: Recommendations for the use of automated gray matter segmentation tools: evidence from Huntington's disease. Front. Neurol. **8**, 519 (2017). https://doi.org/10.3389/fneur.2017.00519
10. Kingma, D.P., Ba, J.L.: Adam: a method for stochastic optimization. In: 3rd International Conference on Learning Representations, ICLR 2015 - Conference Track Proceedings (2015). https://doi.org/10.48550/arxiv.1412.6980
11. Larrazabal, A.J., Nieto, N., Peterson, V., Milone, D.H., Ferrante, E.: Gender imbalance in medical imaging datasets produces biased classifiers for computer-aided diagnosis. Proc. Natl. Acad. Sci. U.S.A. **117**(23), 12592–12594 (2020). https://doi.org/10.1073/pnas.1919012117
12. Ledig, C., et al.: Robust whole-brain segmentation: application to traumatic brain injury. Med. Image Anal. **21**(1), 40–58 (2015). https://doi.org/10.1016/j.media.2014.12.003
13. Ledig, C., Schuh, A., Guerrero, R., Heckemann, R.A., Rueckert, D.: Structural brain imaging in Alzheimer's disease and mild cognitive impairment: biomarker analysis and shared morphometry database. Sci. Rep. **8**(1), 11258 (2018). https://doi.org/10.1038/s41598-018-29295-9
14. Mehrabi, N., Morstatter, F., Saxena, N., Lerman, K., Galstyan, A.: A survey on bias and fairness in machine learning. ACM Comput. Surv. **54**(6) (2021). https://doi.org/10.1145/3457607
15. Puyol-Antón, E., et al.: Fairness in cardiac magnetic resonance imaging: assessing sex and racial bias in deep learning-based segmentation. Front. Cardiovascul. Med. **9**, 664 (2022). https://doi.org/10.3389/fcvm.2022.859310
16. Puyol-Antón, E., et al.: Fairness in cardiac MR image analysis: an investigation of bias due to data imbalance in deep learning based segmentation. In: de Bruijne, M., et al. (eds.) MICCAI 2021. LNCS, vol. 12903, pp. 413–423. Springer, Cham (2021). https://doi.org/10.1007/978-3-030-87199-4_39
17. Roy, A.G., Conjeti, S., Sheet, D., Katouzian, A., Navab, N., Wachinger, C.: Error corrective boosting for learning fully convolutional networks with limited data. In: Descoteaux, M., Maier-Hein, L., Franz, A., Jannin, P., Collins, D.L., Duchesne, S. (eds.) MICCAI 2017. LNCS, vol. 10435, pp. 231–239. Springer, Cham (2017). https://doi.org/10.1007/978-3-319-66179-7_27
18. Ruijsink, B., et al.: Fully automated, quality-controlled cardiac analysis from CMR: validation and large-scale application to characterize cardiac function. JACC: Cardiovascul. Imaging **13**(3), 684–695 (2020). https://doi.org/10.1016/j.jcmg.2019.05.030
19. Sudre, C.H., Li, W., Vercauteren, T., Ourselin, S., Jorge Cardoso, M.: Generalised dice overlap as a deep learning loss function for highly unbalanced segmentations. In: Cardoso, M.J., et al. (eds.) DLMIA/ML-CDS -2017. LNCS, vol. 10553, pp. 240–248. Springer, Cham (2017). https://doi.org/10.1007/978-3-319-67558-9_28

Volume is All You Need: Improving Multi-task Multiple Instance Learning for WMH Segmentation and Severity Estimation

Wooseok Jung[1], Chong Hyun Suh[2(✉)], Woo Hyun Shim[2], Jinyoung Kim[1], Dongsoo Lee[1], Changhyun Park[1], Seo Taek Kong[1], Kyu-Hwan Jung[1], Hwon Heo[3], and Sang Joon Kim[2]

[1] VUNO Inc., Seoul, South Korea
wooseok.jung@vuno.co

[2] Department of Radiology and Research Institute of Radiology, Asan Medical Center, University of Ulsan College of Medicine, Seoul, South Korea
chonghyunsuh@amc.seoul.kr

[3] Department of Convergence Medicine, Asan Medical Center, University of Ulsan College of Medicine, Seoul, South Korea
https://www.vuno.co/en/

Abstract. White matter hyperintensities (WMHs) are lesions with unusually high intensity detected in T2 fluid-attenuated inversion recovery (T2-FLAIR) MRI images, commonly attributed to vascular dementia (VaD) and chronic small vessel ischaemia. The Fazekas scale is a measure of WMH severity, widely used in radiology research. Although stand-alone WMH segmentation methods have been extensively investigated, a model encapsulating both WMH segmentation and Fazekas scale prediction has not. We propose a novel multi-task multiple instance learning (MTMIL) model for simultaneous WMH lesions segmentation and Fazekas scale estimation. The model is initially trained only for the segmentation task to overcome the difficulty of the manual annotation process. Afterwards, volume-guided attention (VGA) obtained directly from instance-level segmentation results figure out key instances for the classification task. We trained the model with 558 in-house brain MRI data, where only 58 of them have WMH annotations. Our MTMIL method reinforced by segmentation results outperforms other multiple instance learning methods.

Keywords: Multi-task multiple instance learning · Volume guided attention · WMH segmentation · Fazekas scale prediction

1 Introduction

White matter hyperintensity (WMH) lesions manifest increased brightness patterns in T2 fluid-attenuated inversion recovery (T2-FLAIR) imaging. These stimuli are usually associated with chronic small vessel ischaemia or myelin

A. Abdulkadir et al. (Eds.): MLCN 2022, LNCS 13596, pp. 23–31, 2022.
https://doi.org/10.1007/978-3-031-17899-3_3

loss, which might lead to disorders including cognitive impairment and vascular dementia [4,15]. Despite its clinical importance, annotating WMH is time-consuming and can be done only by practiced raters. The Fazekas scale shown in Fig. 1(a) quantifies the amount of WMH as the size of lesions and their conjunctions [3], but manually scoring WMH may not reflect WMH burdens precisely and lead to high inter-observer variability. In response, segmentation algorithms based on deep-learning have been recently developed, including those submitted to the 2017 MICCAI WMH segmentation challenge [10]. Lead algorithms in the challenge were based on the 2D U-Net [17] architecture differing in loss functions [14] used for training or post-processing methods [11].

Multi-task learning algorithms can segment regions of interest while diagnosing or classifying an input image [1,5,19,21]. Many of them handling MRI exploit 3D encoding backbone; however, they are computationally demanding and known to under-perform efficient 2D models on WMH segmentation [10,11,14]. This work, therefore, concerns WMH segmentation and Fazekas scale classification using 2D axial images.

However, Fazekas scales are deduced from all lesions in the 3D T2-FLAIR image, and therefore we bring multiple instance learning (MIL) into our task. MIL is a weakly-supervised method where a label corresponding to a group of instances instead of each individual is available [2]. A full 3D MRI image is interpreted as a group of 2D axial slices and assigned a "bag" Fazekas scale label accordingly. Regarding the strong correlation between the Fazekas scale and WMH segmentation result (Fig. 1(b)), we propose a refinement to attention-based MIL [6], where the slice-wise attention vector is inherited from segmented WMH volume of each slice instead of learning the vector. To further overcome the shortage of labelled data, we initially pre-train the model with segmentation data, followed by adding the classification task.

This work establishes a multi-task multiple instance learning (MTMIL) for segmentation of WMH lesions and classification of Fazekas scales working with a small number of segmentation annotations. The main challenge in classifying Fazekas scales from a 2D model is that the Fazekas scale is a global assessment of all lesions that is not accessible to a 2D classification model. We propose a MIMTIL with volume-guided attention (VGA) algorithm that overcomes the aforementioned challenges. Each component of our method shows enhanced performance while retaining a low computational footprint with slice-level classifiers.

2 Materials and Methods

In this section, we first introduce a semi-supervised multi-task model to handle limited labelled data. Then, we describe multiple instance learning for prediction, which exploits segmentation outputs directly. An overall scheme of our proposed method is presented in Fig. 2.

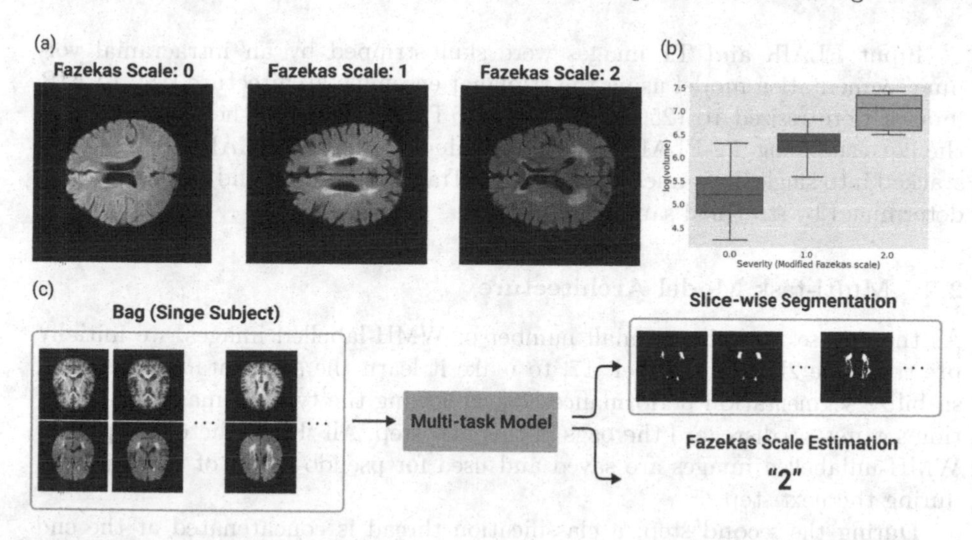

Fig. 1. (a) T2-FLAIR axial slices and corresponding Fazekas scales. (b) Boxplot illustrates the relationship between slice-wise WMH volume and Fazekas scale. (c) The multi-task problem considered in this work.

2.1 Dataset Description and Preprocessing

An in-house dataset containing axial T2-FLAIR and T1 MRI images from 800 consecutive patients who visited a tertiary hospital memory clinic due to cognitive impairment was used for this work with Institutional Review Board approvals. Three hundred FLAIR images were reviewed by two experienced neuro-radiologists, and the final WMH segmentation labels were produced with their consensus. All FLAIR images were then categorized into clinically normal, mild, moderate, or severe according to the Fazekas scale [3]. The first two groups were combined into a single class because their distinction is not clinically significant (i.e., (normal, mild, moderate, severe) = (0,0,1,2)) [8,13]. Originally, WMH is divided into two subtypes—deep and periventricular—so as their corresponding Fazekas scale. However, we consider a maximum of two scales per image for simplicity (Table 1).

Table 1. Demographic statistics and number of images labeled, unlabeled, and corresponding to each Fazekas scale.

Category	Train	Validation	Test	Total
# Labeled/unlabeled images	58/500	64/0	178/0	300/500
Age	69.3 ± 10.2	70.3 ± 10.3	68.2 ± 10.8	69.1 ± 10.4
Sex F/M	337/221	40/24	116/62	493/307
# Fazekas scales (0/1/2)	372/128/58	43/12/9	125/36/17	540/176/84

Input FLAIR and T1 images were skull-stripped by an intracranial volume segmentation model using the triplanar ensemble architecture [18]. FLAIR images were resized to (256, 256, 35), and T1 images were then registered to the corresponding T2-FLAIR space. Matching T1 and T2-FLAIR images were stacked into single 2-channel input images. Train, validation, and test splits were determined by stratified sampling.

2.2 Multi-task Model Architecture

As the dataset contains a small number of WMH-labelled images, we initially pre-train the 2D UNet model [17] to make it learn the segmentation task and stabilize segmentation performance [20]. Following the typical image segmentation setting, we disregard the bags in this first step. All the predicted outputs of WMH-unlabelled images are saved and used for pseudo-labels of those images during the next step.

During the second step, a classification thread is concatenated at the end of the UNet encoder to achieve a multi-task scheme [1,5,19,21], but the model differs in how it uses 2D images instead of 3D inputs. The classification thread takes a high-level feature vector as its input (Fig. 2). Global average pooling (GAP) is first applied to reduce the spatial dimension of the input, followed by a multilayer perceptron (MLP), consisting of fully-connected layers, ReLU activation, and dropout layers. Meanwhile, the segmentation thread computes the soft dice loss [7] ($\mathcal{L}_{seg}$ in Fig. 2) between outputs $\hat{\mathcal{Y}}_{mask}$ and pseudo-labels or ground truths $\mathcal{Y}_{mask}$. Finally, mean squared error (MSE) loss $\mathcal{L}_{clf}$ is computed between the output $\hat{\mathcal{Y}}_{bag}$ of the final classification layer and the corresponding true bag label $\mathcal{Y}_{bag}$. Hence, the total loss function $\mathcal{L}_{total}$ becomes

$$\mathcal{L}_{total} = \mathcal{L}_{clf} + \mathcal{L}_{seg}. \tag{1}$$

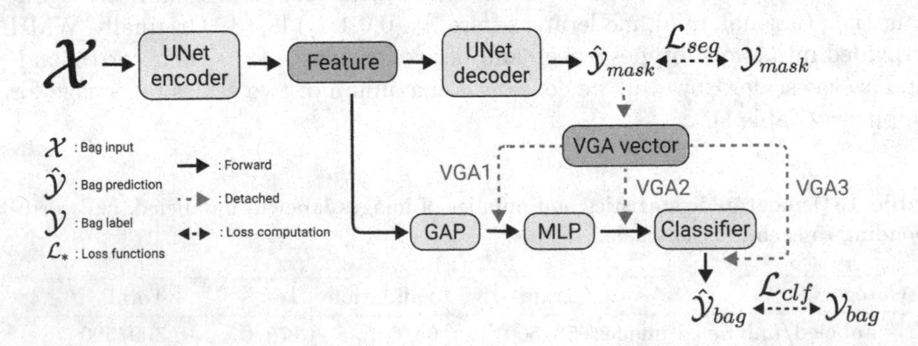

Fig. 2. An overview of the proposed segmentation-guided multi-task multiple instance learning model. If applicable, volume-guided attention reaches out from the segmentation output in detached mode.

2.3 Segmentation-Guided Multiple Instance Learning

Fazekas scale quantification is generally derived from a global assessment of 3D images, but radiologists determine the scale from slices with large WMH. This motivates multiple instance learning, which admits labelled bags of instances whose individual labels are missing. Given a bag $\mathcal{X} = (x_1, \ldots, x_N)$ of size N, containing the 2-channel input MRI images derived from a single subject, the multi-task model returns a segmentation mask $\hat{\mathcal{Y}}_{mask} = (\hat{y}_1^{mask}, \ldots, \hat{y}_N^{mask})$ and an instance-level feature vector $\mathbf{F} = (\hat{x}_1, \ldots, \hat{x}_N)$ for classification. However, each batch has a single bag label not as many as its size. A basic MTMIL model can average the instance predictions to make a bag prediction $\hat{\mathcal{Y}}_{bag}$. Alternatively, one can apply self-attention algorithm and take dot product of the attention vector and the features [6].

Our objective is to enhance MIL performance and complement its uniform attribution to slice-level predictions; the latter is critical when slides do not contain any region of interest. Slice-wise WMH volume is closely related to the severity to which it belongs. If a bag contains slices with large or small volumes, only the larger slice becomes a key instance in bag prediction. However, if all instances have small WMH volumes, they equally contribute to estimating the severity. This motivates us to explore the necessity of instance-wise attention derived from segmentation results.

Volume-Guided Attention (VGA). Attention-based MIL models are typically fortified by additional attention blocks [6,16]. Instead, we obtain a volume-guided attention (VGA) vector $\mathbf{a} = (a_1, \ldots, a_N)$, adjusted from the self-attention algorithm in MIL, which is directly determined from outputs of the segmentation thread. We evaluate WMH volumes $\mathbf{v} = (v_1, \ldots, v_N)$ from $\hat{\mathcal{Y}}_{mask}$, obtained by counting the number of nonzero pixels in $\hat{y}_i^{mask}$ of each instance x_i, to estimate the attention vector $\mathbf{a}$. Therefore $\mathbf{a}$ can be computed as follows:

$$\mathbf{a} = \text{softmax}(\frac{\mathbf{v}}{\tau}), \tag{2}$$

where τ refers to temperature. Small τ attenuates features from slices where WMH are segmented less, whereas large τ balances effects from the all slices in a bag. Let the instance-level feature vector be $\mathbf{F}$. As $\mathbf{a}$ has shape $1 \times N$, the resulting feature vector $\mathbf{F}' := \mathbf{aF}$ can be treated with batch size equals to 1.

Furthermore, we divided the types of VGA by which stage the attention vector is applied. If VGA precedes the MLP block (VGA1), aggregated features are forwarded into the classification thread. On the other hand, if the attention is applied at the end of the MLP block (VGA2), the block will regard instances independently, and only the final classification layer takes the aggregated feature into account. Finally, the attention can be applied after the classification layer (VGA3).

3 Experiments

3.1 Implementation Details

For online preprocessing, we first normalized each channel of all images with mean = 0.456 and std = 0.224 to conform intensity distribution among the data. Flipping horizontal or vertical direction, affine transforms together with translation ratio range = (± 0.1), rotation range = ($\pm 15°$), scaling ratio range = (1 ± 0.1), and Gaussian blur were applied with probability 0.5. Finally, all input images during training are cropped to the size of (2, 200, 200) [14]. We applied the same augmentation module at the multi-task stage, where instances in the same bag can be augmented differently.

We selected a batch size of 16 for the pre-training, and batches were loaded across patients. We fixed the bag size to 16 at the multi-task step but changed the batch size to 1, producing a bag containing instances sampled from axial index 14 to 30, where WMH usually appears.

We used Pytorch 1.7.1 for the implementation of the proposed method. All experiments were tested on two GeForce GTX 1080 Ti and two GeForce GTX TITAN X GPUs with CUDA version 11.2. The learning rate is set to 0.001 initially, but cosine annealing scheduler [12] with minimal learning rate = 1e-6 gradually reduces the learning rate while training. We used ADAM optimizer [9] with ($\beta_1 = 0.9, \beta_2 = 0.999$). We trained models with maximum epochs of 200. Early stopping regularization monitoring validation loss was applied to prevent overfitting.

3.2 Results

Table 2. Segmentation and prediction performances of the WMH segmentation and severity estimation task. Dice score and average volume difference metrics were used to evaluate segmentation. Classification performance was evaluated by accuracy, AUROC, precision, and recall, which were macro-averaged except the accuracy. 0.5 cut-off were applied during the bag prediction. τ is fixed to 0.75 within the VGA family.

Model	Dice score	AVD (ml)	Accuracy	AUROC	Precision	Recall
Segmentation only	**0.877**	**0.313**	N/A	N/A	N/A	N/A
MIL	N/A	N/A	0.635	0.758	0.659	0.588
MIL + SA	N/A	N/A	0.697	0.423	0.234	0.331
MTMIL	0.855	0.410	0.702	0.127	0.234	0.333
MTMIL + SA	0.866	0.505	0.702	0.164	0.234	0.333
MTMIL + VGA1	0.862	0.447	0.787	0.921	0.453	0.505
MTMIL + VGA2	0.855	0.753	**0.831**	**0.931**	**0.779**	**0.723**
MTMIL + VGA3	0.860	0.524	0.820	0.925	0.716	0.675

Before evaluating the proposed model, we examined the performances of several baseline methods without an attention algorithm or with self-attention

[6]. Also, the segmentation performance of the pre-trained model (Segmentation only) is measured to be compared with other multi-task methods. Table 2 shows that pre-training improves classification accuracy for both MIL and MTMIL methods. However, the pre-trained model outperforms the other baselines in segmentation.

We further examined the effects of three types of volume-guided attention mixed with MTMIL. MTMIL models showed superior classification performance than the baseline models if merged with any type of volume-guided attention. Among them, multiplying the attention vector to the MLP block output (VGA2) showed the highest in all classification metrics. We assert that VGA1 might compress excessively at the beginning of classification. VGA3 performs better than VGA1, but passing unweighted features until the end might compromise the final classification layer. Despite the promising classification performance of MTMIL with VGA, segmentation performance is inferior to the classification.

Table 3. Ablation study of VGA subtypes and the temperature τ.

	VGA1	VGA1	VGA1	VGA1	VGA2	VGA2
	0.25	0.5	0.75	1.0	0.25	0.5
Accuracy	0.787	0.607	0.787	**0.831**	0.798	0.798
AUROC	0.920	0.820	0.921	0.927	0.921	0.898
Precision	0.687	0.529	0.453	0.735	0.705	0.714
Recall	0.551	0.620	0.505	**0.751**	0.579	0.654
	VGA2	VGA2	VGA3	VGA3	VGA3	VGA3
	0.75	1.0	0.25	0.5	0.75	1.0
Accuracy	**0.831**	0.775	0.803	0.820	0.820	0.770
AUROC	**0.931**	0.885	0.920	0.926	0.925	0.891
Precision	**0.779**	0.661	0.685	0.754	0.716	0.679
Recall	0.723	0.734	0.728	0.738	0.675	0.703

Ablation study in Table 3 shows that small temperature in VGA function is not preferable. High classification performance is achieved when the temperature is 1 or 0.75. This result implies that giving overwhelming weight to images with large WMH can overlook the other slices. Interestingly, this effect is reduced in VGA3. Unlike VGA1 or VGA2, the attention vector is multiplied at the end of the neural network. It makes slices with large WMH more likely to contribute to the final prediction.

Figure 3 illustrates how the severity prediction is manifested in the proposed framework. Instances with large WMH volumes gain more attention at the classification stage. Greater the attention value, more contribution of corresponding features to the prediction result. If global severity is low, the WMH volumes will be equally small, and slices will have even attention. On the other hand, if

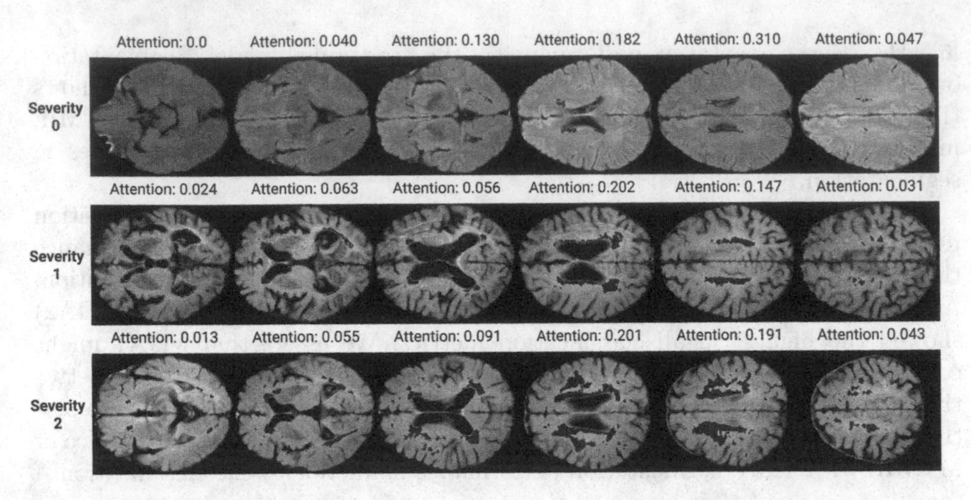

Fig. 3. WMH Segmentation maps of samples with different severity estimate from the model trained with the VGA. Value above each slice denotes the attention value contributed from the slice.

global severity is high, only the slices with a large WMH volume mainly contribute to determine the severity. In this case, the model will like to only take the features of those slices into account for the prediction. The volume-guided attention abates unnecessary features and accentuates features more closely related to its prediction.

4 Conclusion

In this work, we propose a novel multi-task multiple instance learning method. The model fetches parameter-free volume-guided attention to find key instances. Experimental results show that the proposed method outperforms an existing 3D multi-task method in WMH segmentation with similar Fazekas scale prediction performance. The competency of our proposed method comes from its versatility. VGA can be used for various multi-task learning problems in medical image analysis if segmentation and classification results are highly correlated. In future research, we will develop a MIMTL model that differentiates deep and periventricular WMHs.

References

1. Chen, C., Bai, W., Rueckert, D.: Multi-task learning for left atrial segmentation on GE-MRI. In: Pop, M., et al. (eds.) STACOM 2018. LNCS, vol. 11395, pp. 292–301. Springer, Cham (2019). https://doi.org/10.1007/978-3-030-12029-0_32
2. Dietterich, T.G., Lathrop, R.H., Lozano-Pérez, T.: Solving the multiple instance problem with axis-parallel rectangles. Artif. intell. **89**(1–2), 31–71 (1997)

3. Fazekas, F., Chawluk, J.B., Alavi, A., Hurtig, H.I., Zimmerman, R.A.: MR signal abnormalities at 1.5 t in Alzheimer's dementia and normal aging. Am. J. Neuroradiol. **8**(3), 421–426 (1987)
4. Haller, S., et al.: Do brain t2/flair white matter hyperintensities correspond to myelin loss in normal aging? a radiologic-neuropathologic correlation study. Acta Neuropathol. Commun. **1**(1), 14–14 (2013)
5. Hernández, M.d.C.V., et al.: Close correlation between quantitative and qualitative assessments of white matter lesions. Neuroepidemiology **40**(1), 13–22 (2013)
6. Ilse, M., Tomczak, J., Welling, M.: Attention-based deep multiple instance learning. In: International Conference on Machine Learning, pp. 2127–2136. PMLR (2018)
7. Jadon, S.: A survey of loss functions for semantic segmentation. In: 2020 IEEE Conference on Computational Intelligence in Bioinformatics and Computational Biology (CIBCB), pp. 1–7. IEEE (2020)
8. Kim, S., et al.: Periventricular white matter hyperintensities and the risk of dementia: a credos study. Int. Psychogeriat. **27**(12), 2069–2077 (2015)
9. Kingma, D.P., Ba, J.: Adam: a method for stochastic optimization. arXiv preprint arXiv:1412.6980 (2014)
10. Kuijf, H.J., et al.: Standardized assessment of automatic segmentation of white matter hyperintensities and results of the WMH segmentation challenge. IEEE Trans. Med. Imaging **38**(11), 2556–2568 (2019)
11. Li, H., et al.: Fully convolutional network ensembles for white matter hyperintensities segmentation in MR images. NeuroImage **183**, 650–665 (2018)
12. Loshchilov, I., Hutter, F.: SGDR: stochastic gradient descent with warm restarts. arXiv preprint arXiv:1608.03983 (2016)
13. Noh, Y., et al.: A new classification system for ischemia using a combination of deep and periventricular white matter hyperintensities. J. Stroke Cerebrovascul. Dis. **23**(4), 636–642 (2014)
14. Park, G., Hong, J., Duffy, B.A., Lee, J.M., Kim, H.: White matter hyperintensities segmentation using the ensemble u-net with multi-scale highlighting foregrounds. NeuroImage **237**, 118140 (2021)
15. Prins, N.D., Scheltens, P.: White matter hyperintensities, cognitive impairment and dementia: an update. Nat. Rev. Neurol. **11**(3), 157–165 (2015)
16. Qaiser, T., et al.: Multiple instance learning with auxiliary task weighting for multiple myeloma classification. In: de Bruijne, M., et al. (eds.) MICCAI 2021. LNCS, vol. 12907, pp. 786–796. Springer, Cham (2021). https://doi.org/10.1007/978-3-030-87234-2_74
17. Ronneberger, O., Fischer, P., Brox, T.: U-net: convolutional networks for biomedical image segmentation (2015)
18. Sundaresan, V., Zamboni, G., Rothwell, P.M., Jenkinson, M., Griffanti, L.: Triplanar ensemble u-net model for white matter hyperintensities segmentation on MR images. Med. Image Anal. **73**, 102184 (2021)
19. Yang, J., Hu, J., Li, Y., Liu, H., Li, Y.: Joint PVL detection and manual ability classification using semi-supervised multi-task learning. In: de Bruijne, M., et al. (eds.) MICCAI 2021. LNCS, vol. 12907, pp. 453–463. Springer, Cham (2021). https://doi.org/10.1007/978-3-030-87234-2_43
20. Zhang, W., et al.: Deep model based transfer and multi-task learning for biological image analysis. IEEE Trans. Big Data **6**(2), 322–333 (2016)
21. Zhou, Y., et al.: Multi-task learning for segmentation and classification of tumors in 3D automated breast ultrasound images. Med. Image Anal. **70**, 101918 (2021)

Self-supervised Test-Time Adaptation for Medical Image Segmentation

Hao Li[1(✉)], Han Liu[1], Dewei Hu[1], Jiacheng Wang[1], Hans Johnson[2], Omar Sherbini[4], Francesco Gavazzi[4], Russell D'Aiello[4], Adeline Vanderver[4], Jeffrey Long[2], Jane Paulsen[3], and Ipek Oguz[1]

[1] Vanderbilt University, Nashville, USA
hao.li.1@vanderbilt.edu
[2] University of Iowa, Iowa, USA
[3] University of Wisconsin, Madison, USA
[4] Children's Hospital of Philadelphia, Philadelphia, USA

Abstract. The performance of convolutional neural networks (CNNs) often drop when they encounter a domain shift. Recently, unsupervised domain adaptation (UDA) and domain generalization (DG) techniques have been proposed to solve this problem. However, access to source domain data is required for UDA and DG approaches, which may not always be available in practice due to data privacy. In this paper, we propose a novel test-time adaptation framework for volumetric medical image segmentation without any source domain data for adaptation and target domain data for offline training. Specifically, our proposed framework only needs pre-trained CNNs in the source domain, and the target image itself. Our method aligns the target image on both image and latent feature levels to source domain during the test-time. There are three parts in our proposed framework: (1) multi-task segmentation network (Seg), (2) autoencorders (AEs) and (3) translation network (T). Seg and AEs are pre-trained with source domain data. At test-time, the weights of these pre-trained CNNs (decoders of Seg and AEs) are fixed, and T is trained to align the target image to source domain at image-level by the autoencoders which optimize the similarity between input and reconstructed output. The encoder of Seg is also updated to increase the domain generalizability of the model towards the source domain at the feature level with self-supervised tasks. We evaluate our method on healthy controls, adult Huntington's disease (HD) patients and pediatric Aicardi Goutières Syndrome (AGS) patients, with different scanners and MRI protocols. The results indicate that our proposed method improves the performance of CNNs in the presence of domain shift at test-time.

Keywords: Self-supervised · Test-time training · Test-time adaptation · Segmentation

---

Supplementary Information The online version contains supplementary material available at https://doi.org/10.1007/978-3-031-17899-3_4.

A. Abdulkadir et al. (Eds.): MLCN 2022, LNCS 13596, pp. 32–41, 2022.
https://doi.org/10.1007/978-3-031-17899-3_4

1 Introduction

Convolutional neural networks (CNNs) show excellent performance in supervised medical image segmentation tasks if the distribution of the training set (source domain) is tightly matched to the test set (target domain). However, for multi-site studies, domain shift is often present among different imaging sites due to different scanners and MRI protocols. In such scenarios, data from the target domain can be considered as out-of-distribution for the source domain, and the CNN performance can significantly drop during testing due to this domain shift.

Unsupervised domain adaptation (UDA) is a solution to minimize the gap between source and target domains. [2,5,8,13]. However, the UDA normally requires data from both source and target domains to train. Moreover, for multiple target domains, UDA needs to train a separate model for each target domain, which is time-consuming. Another solution is domain generalization (DG), which tries to increase the model generalizability to unseen target domain data [1,4,19]. DG might need large amounts of source domain data or augmented data for training, and it may not adequately represent the data in the unseen target domains to produce robust segmentations. Furthermore, source domain data could be unavailable to researchers/clinicians between sites due to privacy issues. In contrast, a pre-trained model from the source domain is often easier to obtain, but domain shifts could lead to unreliable segmentations when such pre-trained models are directly applied on unseen target domain data.

To produce robust results with access to only the pre-trained models from the source domain and unseen test data, test time adaptation (TTA) could reduce the effects of domain shift by adapting the target data to source data at either the image level or the latent feature level. Wang et al. [18] proposed an image-specific fine tuning pipeline in the testing phase for interactive segmentation by adapting the pre-trained CNN to the unseen target data, and the priors on the predicted segmentations were used for adaptation. Sun et al. proposed a test-time training approach for improving the model performance when domain shift is present between training and test data [15]. They adapt part of the model using a self-supervised rotation task on target data. Furthermore, Wang et al. proposed test-time entropy minimization for adaptation [17]. He et al. proposed a TTA network which is based on autoencoders trained on source domain [7]. During inference, the adaptation is applied on each target data by minimizing the reconstruction loss of autoencoders with fixed weights. Similarly, Karani et al. proposed an adaptable network for TTA [9]. In their work, the weights of the pre-trained segmentation network and the denoising autoencoder are fixed while updating the parameters of the normalized network to achieve adaptation during test-time. However, most TTA methods adapt target data either in image-space or fully/partially in feature-space, and may not have the ability to deal with images with bigger domain shifts, such as anatomical content shifts in addition to image intensity or contrast shifts. In addition, user interaction is needed in [18], which is problematic for large studies. Only feature-level adaptation [15,17] may fail on some cases without image-level adaptation. Finally, a good alignment

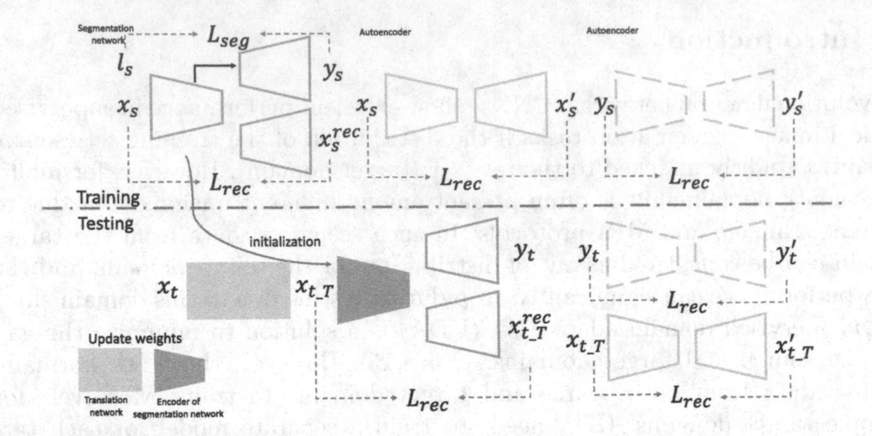

Fig. 1. Proposed test-time adaptation framework. During offline training (top), the multi-task segmentation network (green) is supervised by segmentation and reconstruction losses, and autoencoders (orange) are trained for measuring the similarities. During test-time (bottom), the target image first goes through the translation network (brown), then fed to the pre-trained segmentation network. Only the translation network and the encoder of segmentation network (painted) are updating parameters during testing. $\mathcal{L}_{seg}$ and $\mathcal{L}_{rec}$ denote the Dice loss and MSE loss, respectively. (Color figure online)

between target and source domains may not be possible when only partially features are adapted during test-time [7] or without feature-level adaptation [9].

In this work, inspired by previous works [7,15], we propose a test-time adaptation framework for volumetric medical image segmentation, by adapting the target image at both image and feature levels. Our network has three components: (1) a multi-task segmentation network (Seg) with segmentation and reconstruction tasks, (2) autoencoders (AEs) optimizing the similarity between their input and output, and (3) an image translation network (T) to translate the image from target domain to source domain. The Seg and AEs are trained offline on labeled source data. At test-time, these pre-trained CNNs are fixed, and only the T and the encoder of Seg update weights with target data to achieve test-time adaptation by minimizing the reconstruction losses from self-supervised tasks. We evaluate our method with healthy adults, adult Huntington's disease patients [11] and pediatric Aicardi Goutières Syndrome (AGS) patients [16] for the brain extraction task. The data thus includes different brain sizes, shapes and tissue contrast, and ranges from healthy to severely atrophied anatomy.

2 Materials and Methods

Figure 1 shows our proposed test-time adaptation framework, which consists of three parts: a multi-task segmentation network (Seg), autoencoders (AEs) and a translation network (T). In the offline training phase (top row of Fig. 1), the Seg (green) is trained in a supervised manner with a dataset from source domain

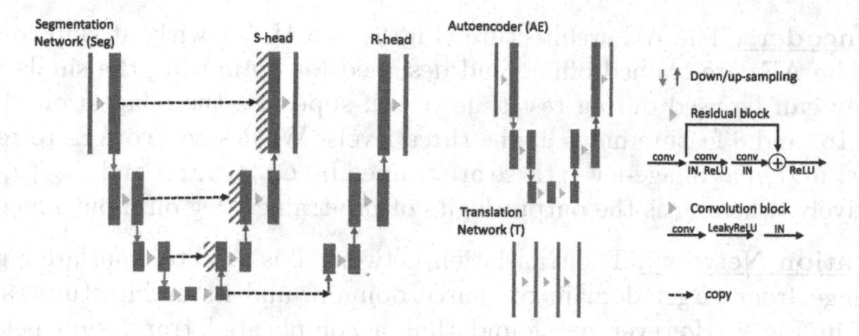

Fig. 2. Network architecture of segmentation network (Seg), Autoencoder (AE) and translation network (T). The S-head and R-head of Seg are for segmentation and reconstruction task, respectively.

$D_s = \{x_s, l_s\}_{s=1}^N$ consisting of input MRIs x_s and corresponding labels l_s. In addition, similar to [7], two AEs (orange) are trained after fixing the weights of Seg to measure the similarities between inputs and reconstructed outputs. At test-time (bottom row), the weights of the AEs and the Seg decoder are fixed. For a given image x_t from target domain, the T (brown) is trained to translate x_t to source domain as image x_{t_T}. Then the translated image x_{t_T} is fed to Seg to obtain the segmentation mask y_t and the reconstructed image $x_{t\_T}^{rec}$. The T and the encoder of Seg are optimized with self-supervised learning, which is the key step in our proposed TTA framework during inference; this is achieved via self-supervised tasks for the reconstruction path of Seg and the AEs. For image-level alignment, the pre-trained AEs control the quality of the translated image x_{t_T}. In other words, the AE loss ($\mathcal{L}_{AE}$) indicates the gap between x_{t_T} and source domain data. Specifically, smaller loss represents the target image x_t has been well translated into source domain. On the other hand, using the fixed decoders, the Seg aims to align the features in feature space, especially for the latent code, to source domain by updating the encoder weights that are self-supervised by the reconstruction path. Thus, the proposed method aligns the target image to source domain on both image and feature levels by updating the weights of T and the encoder of Seg, respectively.

2.1 Networks

Segmentation Network. The segmentation network Seg (Fig. 2) is a multi-task network with segmentation and reconstruction tasks, which is adopted from the 3D U-Net [3] with residual blocks [6]. There are 64 feature maps at each level, except the input and output channel. The whole network takes 3D MRIs as input and outputs the 3D segmentation mask and the reconstructed image. In the offline training phase on $D_s = \{x_s, l_s\}_{s=1}^N$, the network is supervised by a segmentation loss and a reconstruction loss: $\mathcal{L}_S = \mathcal{L}_{seg} + \lambda\mathcal{L}_{rec}$. During test-time, the decoder weights are fixed, and only the encoder weights are updated with self-supervised tasks for adapting the target images at the feature level.

Autoencoders. The AE architecture (Fig. 2) is a U-Net without skip connections. The AEs are trained offline and designed for optimizing the similarities, and they can be used during test-time to self-supervise the adaptation. There are 32, 16, and 8 feature maps at the three levels. We design two AEs to reconstruct x_s and y_s at image-level; these are trained by $\mathcal{L}_{rec}(x_s, x'_s)$ and $\mathcal{L}_{rec}(y_s, y'_s)$, respectively, where y_s is the output logits of pre-trained Seg on input image x_s.

Translation Network. The translation network T is used to translate a given test image from target domain to source domain, and its architecture can be viewed in Fig. 2. However, we found that a complicated translation network would lead to blurry images and geometry shifts, as also discussed by [7,9]. We found that convolution with kernel size 3 also caused similar problems in our experiments. Thus, to preserve the image quality and information, we build T as a shallow network, which consists of three conv-norm-act layers with $1 \times 1 \times 1$ convolution, IN and LeakyReLU activation function for each layer. The channel numbers are 64, 64, and 1, respectively. In this design, the translation network is able to mimic the intensity and contrast for different scanners or imaging protocols without any major changes of geometry. The T takes images from target domain as inputs and produces translated images which are closer to the source domain. During testing, the translation network is optimized by self-supervised tasks for each target image.

2.2 Test-Time Optimization

At test-time, two components have updated weights in our framework: the encoder of Seg and T. To increase the generalizability of Seg to target images in feature space, at test-time, the encoder is initialized with the pre-trained weights and updated for all test images instead of reinitialization after each subject. This allows the encoder of Seg to take advantage of the distributional information of the target dataset. For T, the weights are initialized and updated for adapting each target image. In addition, the translation is self-supervised by AEs during test-time. In this way, the target image is aligned to source domain at the image level. For our experiments, we used a single optimizer to update the weights of encoder of Seg and T rather than updating them separately.

2.3 Datasets and Implementation Details

We evaluate our proposed method on the scenario of moderate domain shift (inside same multi-site dataset) and big domain shift (across two different multi-site datasets of different age groups and diseases) using T1-w MRIs for segmenting whole brain masks (i.e., skullstripping).

Adult Dataset. We use a subset of the multi-site PREDICT-HD database [11], with 3D T1-w 3T MRIs of 16 healthy control subjects (multiple visits per subject, total of 26 MRIs) and 10 Huntington's disease (HD) patients (19 MRIs). The training/validation sets consist of 14/2 healthy control subjects with 22/4 MRIs

Table 1. Quantitative results for HD and AGS datasets. Bold numbers indicate best performance. Significant improvements between the proposed method and all compared methods (2-tailed paired t-test, $p < 0.005$) are denoted via $*$.

	HD dataset			AGS dataset		
Method	Dice	ASD	HD95	Dice	ASD	HD95
NA	96.67(.010)	.442(.385)	1.93(2.25)	90.12(.057)	2.600(1.675)	9.82(4.72)
HM	96.77(.008)	.434(.246)	1.63(1.35)	90.54(.055)	2.222(1.595)	8.08(4.74)
CycleGAN	96.52(.012)	.504(.500)	2.12(2.60)	85.16(.056)	3.630(1.715)	11.4(4.34)
m-NA	96.68(.011)	.509(.451)	2.02(2.53)	90.21(.050)	1.916(1.304)	6.67(3.84)
m-adp	96.13(.012)	.690(.523)	3.82(4.08)	90.63(.048)	1.840(1.245)	6.58(3.85)
SDA-Net [7]	96.55(.009)	.419(.212)	**1.54(0.48)**	90.69(.046)	1.950(1.356)	6.92(4.76)
DAE [9]	96.54(.011)	.462(.372)	1.89(1.99)	90.85(.045)	2.225(1.489)	8.28(5.10)
Proposed	**96.78(.008)**	**.363(.168)**	1.56(0.40)	**92.07(.039)***$*$	**1.154(0.890)**$*$	**4.35(3.06)**$*$

respectively, and all HD subjects are used for testing. The training and validation MRIs are from a single type of scanner, and the testing set are from several other scanners.

Pediatric AGS Dataset. We use a multi-site dataset of 3D T1-w MRIs (1.5T and 3T) from 58 Aicardi Goutières Syndrome (AGS) subjects [16]. These patients range from infants to teenagers. We again use 16/2 MRIs from adult healthy controls (first dataset) for training/validation, and all AGS subjects are used during testing. The preprocessing steps can be found in [10]. Additionally, images were resampled to $96 \times 96 \times 96$.

Implementation Details. The Dice loss [12] is used for segmentation ($\mathcal{L}_{seg}$) during training. In addition, MSE loss was used for every $\mathcal{L}_{rec}$ in both training and inference (test-time). The Adam optimizer with L2 penalty 0.00001, $\beta_1 = 0.9$, $\beta_2 = 0.999$ was used for both training and testing. For offline training, a constant learning rate of 0.0001 was used for the segmentation network and autoencoders, and the learning rate was set to 0.00001 in test-time for updating the weights of T and the encoder of Seg. We validated the performance every epoch during offline training, and we used early stopping if the average validation result did not increase for 20 epochs. Additionally, for each target image, the test-time training was stopped if the loss was greater than the previous iteration. The total epochs were set to 200/10 for training and testing. The training/testing batch size was 1. We implemented the models using an NVIDIA TITAN RTX and PyTorch. Our code is publicly available at https://github.com/HaoLi12345/TTA.

3 Results

We compared the proposed method to the following methods: **(1)** no adaptation (NA), i.e., directly apply the pre-trained model on test set, **(2)** histogram

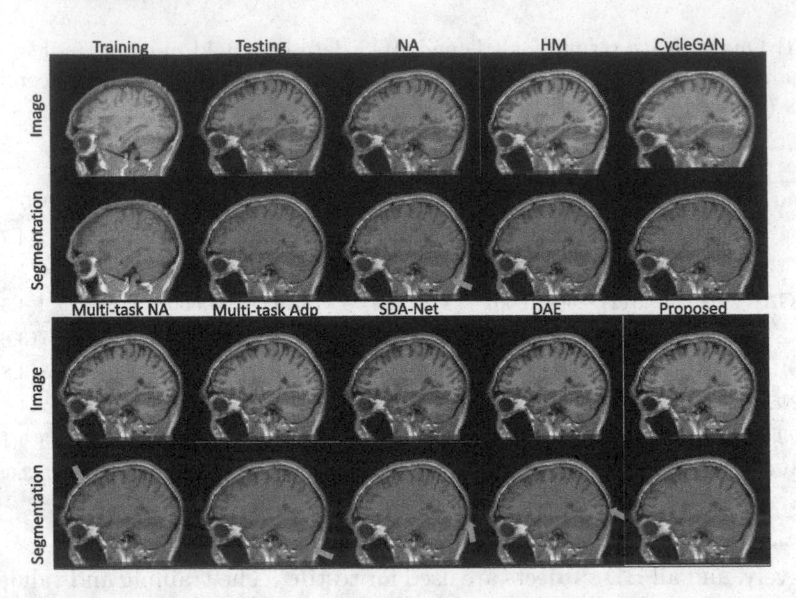

Fig. 3. Adult HD results. $1^{st}/3^{rd}$ rows show the images/adapted images, and $2^{nd}/4^{th}$ rows show segmentation results. Local segmentation defects are highlighted by arrows.

matching (HM) between train and test set [14], **(3)** a typical UDA method that employs CycleGAN [20] to translate the test image to source domain, **(4)** a multi-task network only, without adaptation (m-NA), **(5)** a segmentation network only with adaptation of encoder (m-adp), **(6)** TTA with autoencoders (SDA-Net) [7], and **(7)** TTA with denoised autoencoder (DAE) [9]. For a fair comparison, we use the same network architectures (except the reconstruction path of Seg) and apply identical data augmentations for all methods, except for CycleGAN. Additionally, we modified the SDA-Net to 3D version based on the authors' source code. We use the Dice score, average surface distance (ASD) and 95-percent Hausdorff distance (HD95) for evaluation.

Adult HD Dataset. The quantitative results of the adult HD dataset are shown in the left panel of Table 1. While it uses different scanners than the healthy adult source domain, this dataset has only a moderate domain shift, and may be treated as a supervised segmentation task (NA) in practice. Thus, all methods work relatively well even without adaptation. Nevertheless, our method has the best performance in all metrics except the HD95.

The qualitative results are shown in Fig. 3, where we again observe that all compared methods are able to produce reliable segmentations. However, local defects are present in the baseline methods, as highlighted by orange arrows. Although HM and CycleGAN have better visual quality of adapted image, we note that these methods require access to the source domain data. Among the TTA methods, the proposed method has the best segmentation performance with good quality adapted image.

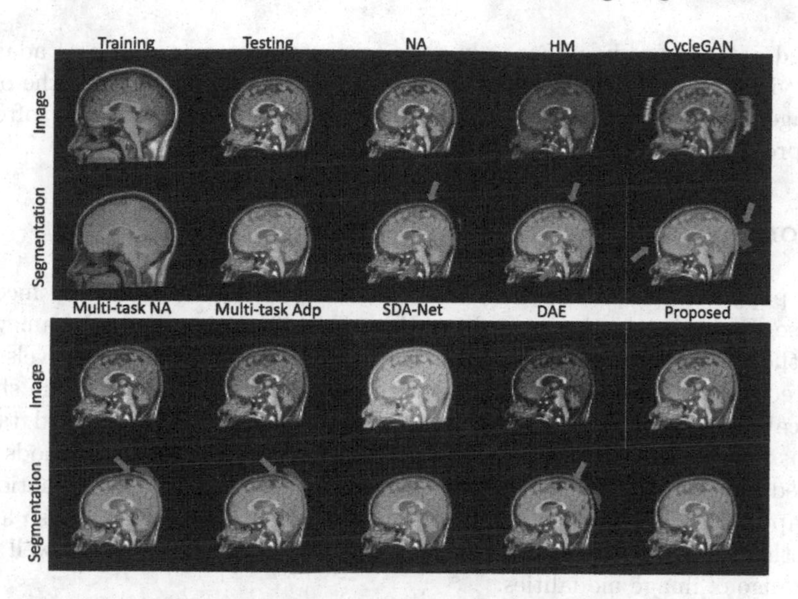

Fig. 4. Pediatric AGS results. Local segmentation defects are highlighted by arrows.

Pediatric AGS Dataset. The quantitative results are shown in the right panel of Table 1. This dataset has more pronounced domain shift from source domain, and our proposed method has the highest Dice score. In addition, while Dice score is not sensitive to local errors, the ASD and HD95 distance metrics demonstrate that our method produces superior segmentations with fewer local defects. 2-tailed paired t-tests show that the proposed method has significantly better ($p < 0.005$) performance for all metrics and compared to each baseline.

Figure 4 shows the qualitative results of adapted images and segmentations. We observe that each baseline method presents either over-segmented or under-segmented regions due to intensity and shape shifts between source and target domains. Specifically, NA and HM under-segment the brain stem area. CycleGAN produces an over-segmented result. For the methods based on multi-task segmentation network, both under-segmented and over-segmented areas appear. Although the SDA-Net produces a plausible segmentation with only small errors in the neck area, it nevertheless achieves worse quantitative results than our proposed method. DAE has good average Dice but poor qualitative results locally (as also evidenced by high ASD and HD95 scores). In contrast, our proposed method produces superior segmentation and is visually closest to the ground truth. We also present the adapted images for each method with image-level adaptation. Even though HM and CycleGAN require training data for image translation, HM changes the contrast in the wrong direction and geometry shifts appear in the CycleGAN since the brain sizes are different between the domains. All TTA methods translate the target image without any major changes and preserve the details of the target image. However, the adapted images from SDA-NET and DAE have visible biases compared to the training image. Our proposed method produces not only a superior segmentation, but also adapted images visually similar to the training data. In addition, compared to the original image, the contrast between cerebrospinal fluid and other tissues is softened in the proposed adapted image, similar to the adult subjects.

4 Conclusion

In this paper, we propose a novel test-time adaptation framework for medical image segmentation in the presence of domain shift. Our proposed framework aligns the target data to source domain at both image and feature levels. We evaluated our method on two datasets with moderate and severe domain shifts. Specifically, intensity and geometry shifts appear between source and target domains for the pediatric AGS dataset. Compared to the baseline methods, our proposed method produced the best segmentations. Quantitative evaluation of the adapted images remains as future work. In future work, we will also apply our method to more datasets with different structures of interest as well as a wider range of image modalities.

Acknowledgements. This work was supported, in part, by NIH grants U01-NS106845 and R01-NS094456. The PREDICT-HD study was funded by the NCATS, the NIH (NS040068, NS105509, NS103475) and CHDI.org.

References

1. Billot, B., Greve, D., Van Leemput, K., Fischl, B., Iglesias, J.E., Dalca, A.V.: A learning strategy for contrast-agnostic MRI segmentation. arXiv preprint arXiv:2003.01995 (2020)
2. Chen, C., et al.: Unsupervised multi-modal style transfer for cardiac MR segmentation. In: Pop, M., et al. (eds.) STACOM 2019. LNCS, vol. 12009, pp. 209–219. Springer, Cham (2020). https://doi.org/10.1007/978-3-030-39074-7_22
3. Çiçek, Ö., Abdulkadir, A., Lienkamp, S.S., Brox, T., Ronneberger, O.: 3D U-Net: learning dense volumetric segmentation from sparse annotation. In: Ourselin, S., Joskowicz, L., Sabuncu, M.R., Unal, G., Wells, W. (eds.) MICCAI 2016. LNCS, vol. 9901, pp. 424–432. Springer, Cham (2016). https://doi.org/10.1007/978-3-319-46723-8_49
4. Dou, Q., Coelho de Castro, D., Kamnitsas, K., Glocker, B.: Domain generalization via model-agnostic learning of semantic features. Adv. Neural Inf. Process. Syst. **32** (2019)
5. Dou, Q., Ouyang, C., Chen, C., Chen, H., Heng, P.A.: Unsupervised cross-modality domain adaptation of convnets for biomedical image segmentations with adversarial loss. arXiv preprint arXiv:1804.10916 (2018)
6. He, K., Zhang, X., Ren, S., Sun, J.: Deep residual learning for image recognition. In: Proceedings of the IEEE Conference on Computer Vision and Pattern Recognition, pp. 770–778 (2016)
7. He, Y., Carass, A., Zuo, L., Dewey, B.E., Prince, J.L.: Autoencoder based self-supervised test-time adaptation for medical image analysis. Med. Image Anal. **72**, 102136 (2021)

8. Huo, Y., Xu, Z., Bao, S., Assad, A., Abramson, R.G., Landman, B.A.: Adversarial synthesis learning enables segmentation without target modality ground truth. In: 2018 IEEE 15th International Symposium on Biomedical Imaging (ISBI 2018), pp. 1217–1220. IEEE (2018)
9. Karani, N., Erdil, E., Chaitanya, K., Konukoglu, E.: Test-time adaptable neural networks for robust medical image segmentation. Med. Image Anal. **68**, 101907 (2021)
10. Li, H., et al.: Human brain extraction with deep learning. In: Medical Imaging 2022: Image Processing, vol. 12032, pp. 369–375. SPIE (2022)
11. Long, J.D., Paulsen, J.S., Investigators, P.H., of the Huntington Study Group, C.: Multivariate prediction of motor diagnosis in Huntington's disease: 12 years of predict-hd. Movem. Disorder. **30**(12), 1664–1672 (2015)
12. Milletari, F., Navab, N., Ahmadi, S.A.: V-net: fully convolutional neural networks for volumetric medical image segmentation. In: 2016 Fourth International Conference on 3D Vision (3DV), pp. 565–571. IEEE (2016)
13. Ouyang, C., Kamnitsas, K., Biffi, C., Duan, J., Rueckert, D.: Data efficient unsupervised domain adaptation for cross-modality image segmentation. In: Shen, D., et al. (eds.) MICCAI 2019. LNCS, vol. 11765, pp. 669–677. Springer, Cham (2019). https://doi.org/10.1007/978-3-030-32245-8_74
14. Reinhold, J.C., Dewey, B.E., Carass, A., Prince, J.L.: Evaluating the impact of intensity normalization on MR image synthesis. In: Medical Imaging 2019: Image Processing, vol. 10949, pp. 890–898. SPIE (2019)
15. Sun, Y., Wang, X., Liu, Z., Miller, J., Efros, A., Hardt, M.: Test-time training with self-supervision for generalization under distribution shifts. In: International Conference on Machine Learning, pp. 9229–9248. PMLR (2020)
16. Vanderver, A., et al.: Early-onset aicardi-goutieres syndrome: magnetic resonance imaging (MRI) pattern recognition. J. Child Neurol. **30**(10), 1343–1348 (2015)
17. Wang, D., Shelhamer, E., Liu, S., Olshausen, B., Darrell, T.: Tent: fully test-time adaptation by entropy minimization. arXiv preprint arXiv:2006.10726 (2020)
18. Wang, G., et al.: Interactive medical image segmentation using deep learning with image-specific fine tuning. IEEE Trans. Med. Imaging **37**(7), 1562–1573 (2018)
19. Zhang, L., et al.: Generalizing deep learning for medical image segmentation to unseen domains via deep stacked transformation. IEEE Trans. Med. Imaging **39**(7), 2531–2540 (2020)
20. Zhu, J.Y., Park, T., Isola, P., Efros, A.A.: Unpaired image-to-image translation using cycle-consistent adversarial networks. In: Proceedings of the IEEE International Conference on Computer Vision, pp. 2223–2232 (2017)

Accurate Hippocampus Segmentation Based on Self-supervised Learning with Fewer Labeled Data

Kassymzhomart Kunanbayev[✉], Donggon Jang, Woojin Jeong, Nahyun Kim, and Dae-Shik Kim

KAIST, 291 Daehak-ro, Yuseong-gu, Daejeon, South Korea
{kkassymzhomart,jdg900,woojin05,nhkim21,daeshik}@kaist.ac.kr

Abstract. Brain MRI-based hippocampus segmentation is considered as an important biomedical method for prevention, early detection, and accurate diagnosis of neurodegenerative disorders like Alzheimer's disease. The recent need for developing accurate as well as robust systems has led to breakthroughs making advantage of deep learning, but requiring significant amounts of labeled data, which, in turn, is costly and hardly obtainable. In this work, we try to address this issue by introducing self-supervised learning for hippocampus segmentation. We devise a new framework, based on the widely known method of Jigsaw puzzle reassembly, in which we first pre-train using one of the unlabeled MRI datasets, and then perform a downstream segmentation training with other labeled datasets. As a result, we found our method to capture local-level features for better learning of anatomical information pertaining to brain MRI images. Experiments with downstream segmentation training show considerable performance gains with self-supervised pre-training over supervised training when compared over multiple label fractions.

Keywords: Hippocampus · Segmentation · Self-supervised learning · Jigsaw puzzle reassembly

1 Introduction

There have been numerous studies linking the occurrence of Alzheimer's disease (AD) and the hippocampal part of the brain [14,21]. More specifically, the changes in the volume and structure of the hippocampus are associated with the level of the progression of AD [9,14]. In this regard, clinical analysis and timely diagnosis of the hippocampus are crucial for prevention and treatment. One of the most widely applied methods is hippocampus segmentation of brain magnetic resonance imaging (MRI), which can yield the necessary analytical information regarding the size and morphology of the hippocampal part of the brain. However, given its small size as well as the uniformity of MRI images, along with the importance of precise segmentation, the job of hippocampus segmentation turns out to be costly, time-consuming and requires highly qualified expertise to perform the task.

A. Abdulkadir et al. (Eds.): MLCN 2022, LNCS 13596, pp. 42–51, 2022.
https://doi.org/10.1007/978-3-031-17899-3_5

In this regard, the research community focused on devising more sophisticated strategies that could introduce methods with higher accuracy and robustness. To this end, machine and deep learning-based techniques have recently succeeded in automating and improving the task of hippocampus segmentation [2,6,16,20]. However, despite the success of these methods in automating and speeding up the task, achieved frameworks remain to require large amounts of MRI images with corresponding label masks, which are expensive and laborious to collect.

To address this issue, in computer vision, self-supervised learning methods have been introduced that avoid large amounts of labeled data by learning representations from unlabeled data via pretext task strategies [5,8]. In the medical image domain as well, there have been various works incorporating self-supervised learning [7,15]. Perhaps, one of the notable self-supervised learning methods is the widely known Jigsaw puzzle reassembly, proposed by [17], in which a network takes tiles of an image one at a time as input and solves the puzzle by predicting a correct spatial arrangement, thus learning the feature mapping of object parts. Experiments suggested that pre-training and then transferring weights for retraining on a downstream task outperforms supervised learning [17]. Taleb *et al.* [24] introduced solving multimodal Jigsaw puzzle for medical imaging by incorporating the Sinkhorn operator and exploiting synthetic images during pre-training. Navarro *et al.* [15] analyzed the self-supervised learning for robustness and generalizability in medical imaging. Although these methods utilize a Jigsaw puzzle, there are no specific previous works on hippocampus segmentation. Moreover, previous works usually tend to use the same datasets both for pre-training and downstream training, while in real-world scenarios the target dataset may not be large enough for pre-training.

In this regard, we further devise a new self-supervised framework for the task of hippocampus segmentation by adopting the Jigsaw puzzle reassembly problem. We selected this method because it explicitly limits the context of the network processing as it takes one tile at a time, hence, this arrangement should be useful to learn the anatomical parts of the brain and especially the features related to the hippocampal part, given its relatively small size in the brain. We first pre-train the model on one of the unlabeled brain MRI datasets and then re-train on a downstream segmentation task with other labeled datasets, by experimenting over various labeled data fractions (from 100% to 10%). Both quantitative and qualitative results show that pre-trained initialization leads to considerable performance gains in hippocampus segmentation.

2 Method

2.1 How Jigsaw Puzzle is Solved

The original implementation of Jigsaw Puzzle reassembly [17] used the context-free architecture, by building a siamese-ennead convolutional network with shared weights based on AlexNet [13]. The image tiles are first randomly permuted so that image patches are reordered, and fed to the network one at a

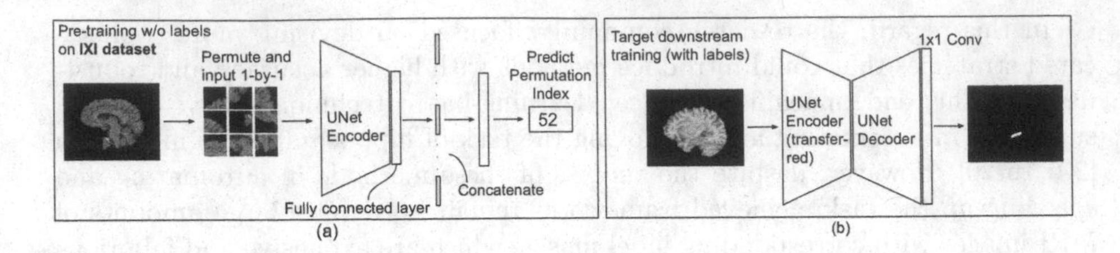

Fig. 1. Self-supervised learning-based (a) pre-training and (b) downstream training framework for hippocampus segmentation.

time. This architecture computes features for each tile separately and then concatenates them to feed as an input to the final fully connected layer (see Fig. 1). Afterwards, the network needs to predict the index of the permutation. The output of the network can be regarded in terms of conditional probability density as follows:

$$p(S|I_1, I_2, ..., I_9) = p(S|F_1, F_2, ..., F_9) \prod_{i=1}^{9} p(F_i|I_i), \tag{1}$$

where S is a permutation configuration, I_i is the i-th image tile, while F_i is the i-th corresponding feature representation after the final fully connected layer. If S is written as a list of positions $S = (L_1, .., L_9)$, then conditional probability distribution would be decomposed into independent terms:

$$p(L_1, L_2, ..., L_9|F_1, F_2, ..., F_9) = \prod_{i=1}^{9} p(L_i|F_i), \tag{2}$$

which implies that each tile position L_i is defined by corresponding F_i [17].

Moreover, one important contribution of Jigsaw puzzle reassembly is how it ensures learning correct representations that will be useful in leveraging the performance during the target downstream training, not just in solving the pretext Jigsaw puzzle task. This is because the network may be prone to learning the latter by following simpler solutions, called *shortcuts* [17]. In the following subsection, we discuss our framework for hippocampus segmentation and how we incorporated the idea of preventing shortcuts during pre-training.

2.2 Framework for Hippocampus Segmentation

Our training framework generally follows the original Jigsaw implementation [17], but with certain adjustments for medical imaging. For example, instead of the AlexNet network, we use the UNet encoder-decoder network [19] for its popularity and remarkable performance in medical imaging, as well as the convenient design. Instead of training the entire network for Jigsaw puzzle reassembly, which could be both time-consuming and computationally expensive, we pretrain only the encoder part and use its weights as initialization for downstream segmentation training. The framework flow is visualized in Fig. 1b.

As for Jigsaw puzzle reassembly, we divide the MRI image into a 3×3 grid to obtain 9 tiles. To further avoid shortcuts, the tiles are randomly cropped to a smaller size so that the model avoids solving the problem by learning edge-related features instead of object positions, so cropping will ensure random shifts in edges [17].

The UNet encoder network followed by fully connected linear layers needs to predict the permutation index. Note that for 9 tiles, there are $9! = 362880$ permutations possible. However, as in [17], to ensure that tiles are shuffled well enough, we selected 1000 permutations based on the Hamming distance. In this way, the network will predict one of the indexes from these 1000 permutations.

Downstream segmentation training is performed by initializing the encoder with pre-trained weights and randomly initializing the decoder. By pre-training only the encoder we ensure that the encoder learns global anatomical features needed to localize the hippocampus, while the decoder will be trained along with encoder weights for accurate hippocampus segmentation. Additionally, 1×1 convolutional layer is added as the last layer to facilitate the segmentation. More details can be seen in Fig. 1.

2.3 Datasets

IXI Dataset. For pre-training, we obtained T1-weighted images of a widely known IXI dataset (https://brain-development.org/ixi-dataset/) which was collected from three different hospitals: Hammersmith (Philips 3T scanner), Guy's (Philips 1.5T scanner), and Institute of Psychiatry (GE 1.5T scanner). There are a total of 579 MRI images, each having 150 slices sized 250×250. To improve the training outcome, the brain MRI volumes were preprocessed in the following order: (i) brain extraction was applied using the Brain Extraction Tool [23] (available as a part of the FMRIB Software Library) in order to remove non-brain areas that could affect the following pre-processing steps; (ii) the intensity inhomogeneity was applied using the N3 package of the MINC toolkit [22]; (iii) min-max normalization was performed volume-wise to normalize the values.

EADC-ADNI HarP Dataset. For downstream training experiments, we utilized publicly available hippocampus segmentation HarP dataset that was collected as a part of the EADC-ADNI Harmonized Protocol project [1,3,10,12,18] from the Alzheimer's disease Neuroimaging Initiative (ADNI) database (https://adni.loni.usc.edu). The dataset contains 135 T1-weighted MRI volumes [18], consisting of 197 slices, each with a size of 189×233, and their released segmentation masks [4]. The same pre-processing steps as in the IXI dataset were applied to this dataset.

Decathlon Dataset. Additionally, we used a hippocampus segmentation dataset from the Medical Segmentation Decathlon challenge [11]. The dataset includes 265 training and 130 test volumes, but we only utilized the training set due to the unavailability of label masks for the test set. The volumes contain cropped parts of the hippocampal area each with various sizes, hence we only selected images with a size of 32×32. No pre-processing was applied.

2.4 Experimental Setup

Pre-training. The pre-training phase was performed using the IXI dataset. To learn the relevant feature representations from the images that include the hippocampus, we discard the MRI images (slices) that do not contain substantial and meaningful brain information and utilize only images that vividly contain the brain anatomical parts. We centrally crop a 225×225 pixel image from the given MRI image and divide it into 9 tiles, each with a size of 75×75. We further crop it to a size of 64×64 to prevent shortcuts and resize the cropped tiles back to 75×75. The training was performed for 70 epochs with a batch size of 64. The stochastic gradient descent optimizer was used as in [17] but with an initial learning rate of 0.001 and a decay rate of 0.1 every 30 epochs.

Downstream Segmentation Training. The dataset split was done subject-wise, so the segmentation training was performed on 90% of volumes of each labeled dataset, while the remaining 10% is set out as a test set for final testing. In order to perform in-depth analysis of the experiments, we conduct comparisons among three training settings: **finetuning**, **linear**, and **random initialization** (baseline). The **finetuning** implies using pre-trained weights of the encoder and training it along with the decoder, while **linear** indicates freezing of the pre-trained weights of the encoder and training only the decoder, and finally, **random initialization** indicates randomly initializing both the encoder and decoder. In addition, to evaluate the performance of the method on both large and small-sized datasets, we performed experiments on 10%, 20%, 50%, and 100% label fractions of the data and reported the results. The segmentation task training was performed for 70 epochs and was repeated 3 times to report the average. The batch size was also selected 64. The quantitative results were evaluated using the Dice coefficient. The Adam optimizer was used with an initial learning rate of 0.001 and a decay rate of 0.1 every 30 epochs.

Network Details. The UNet implementation that we used follows the structure of the original implementation [19], so its encoder part contains five double convolutional layers with 3×3 kernels, as well as a batch normalization and a ReLU activation after each convolutional layer. The max pooling operator with a stride of 2 is used between the double convolutional layers. Thus, the total number of trainable parameters in the UNet encoder is ~ 9.4 million. During the pretraining, there are two fully connected layers following the UNet encoder with output sizes of 1024 and 4096, each of which is followed with a ReLU activation and a dropout layer with a rate of 0.5. There is also the final classification layer with an output of 1000. After the pre-training, these layers are discarded, and only the weights of the encoder are transferred for further re-training. In the decoder part, there are four double convolutional layers but with in-between

Table 1. Hippocampus segmentation results (Dice coefficient, %) on **the HarP** test set. The downstream segmentation training was conducted with different label fractions of the training data set. The **bold** numbers indicate the best results.

	Label fractions			
	100%	50%	20%	10%
Linear (decoder training, ours)	**99.94**	**99.68**	88.50	89.32
Fine-tuning (ours)	99.93	86.00	**97.44**	**93.11**
Random initialization	94.45	94.09	97.22	91.60

Table 2. Hippocampus segmentation results (Dice coefficient, %) on **the Decathlon** test set. The downstream segmentation training was conducted with different label fractions of the training data set. The **bold** numbers indicate the best results.

	Label fractions			
	100%	50%	20%	10%
Linear (decoder training, ours)	**92.37**	90.90	81.84	63.33
Fine-tuning (ours)	87.18	**92.16**	**91.97**	**88.55**
Random initialization	76.60	88.89	64.47	68.50

upsampling operators with a scale factor of 2 and a bilinear transformation algorithm. After the last double convolutional layer of the decoder, to facilitate the segmentation, there is a final 1×1 convolutional layer with an output sigmoid activation.

The training of the framework was conducted on GPU servers with NVIDIA Titan RTX (24GB) and NVIDIA RTX A6000 (48GB). Python-based PyTorch deep learning framework was used for implementation. The implementation code is available at the following link: https://github.com/qasymjomart/ssl_jigsaw_hipposeg.

3 Results

3.1 Quantitative Results

Tables 1 and 2 illustrate the quantitative results comparing the fine-tuning and linear settings with the random initialization on the HarP and Decathlon datasets, respectively. The test results suggest the superior performance of models pre-trained via the proposed method over all label fractions. On the HarP dataset, the linear setting consistently resulted in the highest segmentation accuracy in high label fractions, while the fine-tuning setting demonstrated a higher performance in the case of lower label fractions. These observations explain how pre-trained weights contribute to the overall performance over various amounts

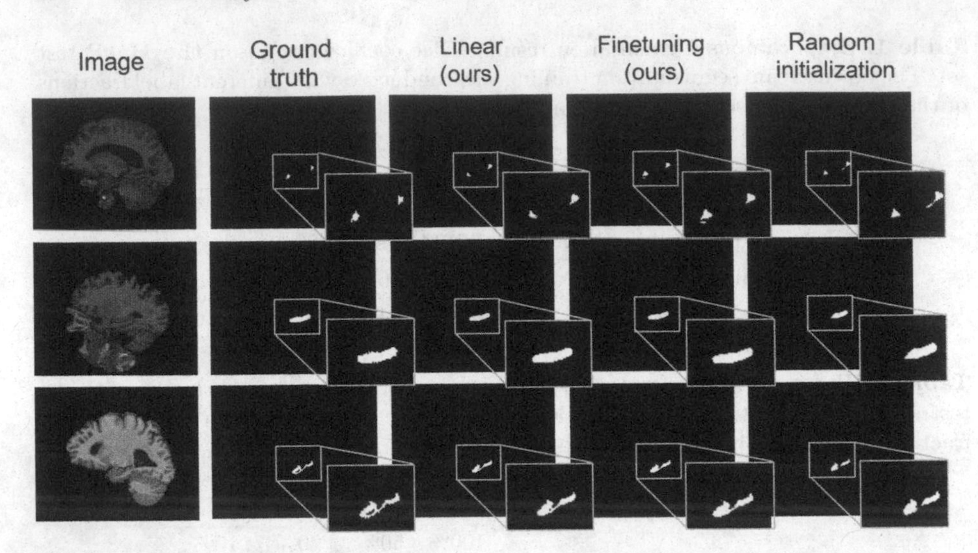

Fig. 2. Randomly selected qualitative results on **the HarP** test set. The results are from the downstream segmentation training with a label fraction of 10%.

of labeled data. Retraining using the pre-trained weights through the fine-tuning setting with fewer labeled data led to better alignment of the encoder and decoder weights in learning segmentation.

Similarly, the experiments on the Decathlon dataset yielded superior performance, where the fine-tuning allowed to achieve considerable performance gains as fewer data is utilized in downstream training. Large differences in accuracy of the 20% and 10% label fractions demonstrate that the fine-tuning setting turned out to be effective in leveraging the segmentation performance.

3.2 Qualitative Results

Qualitative results in Figs. 2 and 3 depict some of the randomly selected MRI brain images with corresponding ground truth and predicted masks. The masks predicted by our method on the HarP dataset exhibited significant similarity with the ground truth. In the first and third rows, the randomly initialized model overpredicted, while in the second row, it underpredicted the hippocampus area.

Similarly, Fig. 3 illustrates clear comparisons in the case of the Decathlon dataset. In all rows, the linear and random initialization demonstrated different predicted shapes and areas, meanwhile, the fine-tuning resulted in more resembling shapes and areas. These qualitative results agree with the quantitative results and suggest that the pre-training via the proposed method provides better capability in predicting the shape and edges of the hippocampus.

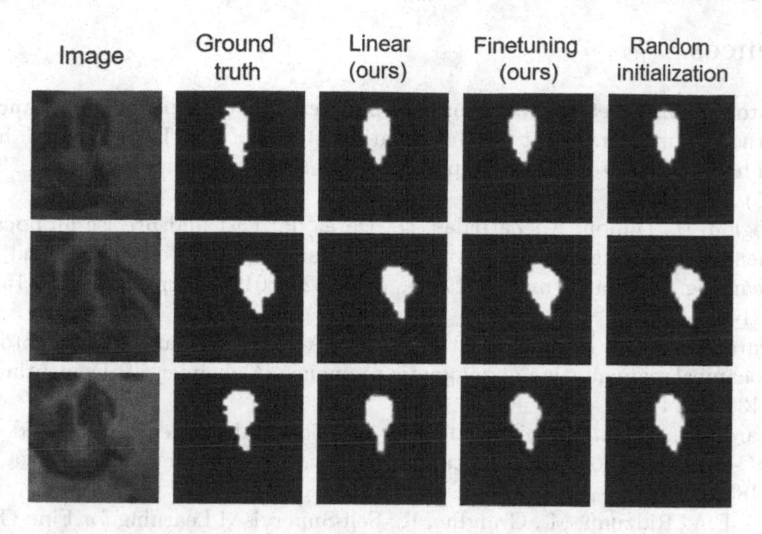

Fig. 3. Randomly selected qualitative results on **the Decathlon** test set. The results are from the downstream segmentation training with a label fraction of 10%.

4 Discussion and Conclusion

In general, results from both datasets indicate that the fine-tuning setting results in better segmentation performance, especially when fewer labeled data is utilized. This is particularly notable when only 10% of data is used for downstream training (see Tables 1 and 2). It is also important to note that for downstream training, we made use of all MRI slices, including those without clear brain anatomical features or even hippocampus, implying empty masks. This imbalance in data may lead to a failure in capturing brain-related features during downstream training. Hence, we conjecture that the effect of pre-trained weights may be better preserved in lower label fractions since they tend to have shorter training. Additionally, the linear settings showed that the pre-trained representations are useful in segmentation. More analysis, including cross-validation, is needed to further address this issue.

Developing accurate as well as robust systems for hippocampus segmentation is a prominent issue for early diagnosis and treatment of AD. In this work, a novel framework that may enhance the overall performance of such systems with fewer amounts of labeled data has been devised. The future research will focus on more in-depth analysis and further development of the framework using other state-of-the-art self-supervised learning methods, and comparing it with other hippocampus segmentation baselines.

Acknowledgements. This work was supported by the Engineering Research Center of Excellence (ERC) Program supported by National Research Foundation (NRF), Korean Ministry of Science & ICT (MSIT) (Grant No. NRF-2017R1A5A1014708).

References

1. Apostolova, L.G., et al.: Relationship between hippocampal atrophy and neuropathology markers: a 7t MRI validation study of the EADC-ADNI harmonized hippocampal segmentation protocol. Alzheimer's & Dementia **11**(2), 139–150 (2015)
2. Ataloglou, D., Dimou, A., Zarpalas, D., Daras, P.: Fast and precise hippocampus segmentation through deep convolutional neural network ensembles and transfer learning. Neuroinformatics **17**(4), 563–582 (2019). https://doi.org/10.1007/s12021-019-09417-y
3. Boccardi, M., et al.: Delphi definition of the EADC-ADNI harmonized protocol for hippocampal segmentation on magnetic resonance. Alzheimer's & Dementia **11**(2), 126–138 (2014)
4. Boccardi, M., et al.: Training labels for hippocampal segmentation based on the EADC-ADNI harmonized hippocampal protocol. Alzheimer's & Dementia **11**(2), 175–183 (2015)
5. Breiki, F.A., Ridzuan, M., Grandhe, R.: Self-Supervised Learning for Fine-Grained Image Classification. arXiv preprint arXiv:2107.13973 (2021)
6. Carmo, D., Silva, B., Yasuda, C., Rittner, L., Lotufo, R.: Hippocampus segmentation on epilepsy and Alzheimer's disease studies with multiple convolutional neural networks. Heliyon **7**(2), e06226 (2021)
7. Chaitanya, K., Erdil, E., Karani, N., Konukoglu, E.: Contrastive learning of global and local features for medical image segmentation with limited annotations (2020). http://hdl.handle.net/20.500.11850/443425
8. Chen, T., Kornblith, S., Norouzi, M., Hinton, G.: A Simple Framework for Contrastive Learning of Visual Representations (2020)
9. Du, A.T.: Magnetic resonance imaging of the entorhinal cortex and hippocampus in mild cognitive impairment and Alzheimer's disease. J. Neurol. Neurosurg. Psychiat. **71**(4), 441–447 (2001)
10. Frisoni, G.B., et al.: The EADC-ADNI harmonized protocol for manual hippocampal segmentation on magnetic resonance: evidence of validity. Alzheimer's & Dementia **11**(2), 111–125 (2014)
11. Isensee, F., et al.: nnU-Net: Self-adapting Framework for U-Net-Based Medical Image Segmentation (2018)
12. Jack, C.R., et al.: The Alzheimer's disease neuroimaging initiative (ADNI): MRI methods. J. Magnet. Reson. Imaging **27**(4), 685–691 (2008)
13. Krizhevsky, A., Sutskever, I., Hinton, G.E.: ImageNet classification with deep convolutional neural networks. In: Pereira, F., Burges, C., Bottou, L., Weinberger, K. (eds.) Advances in Neural Information Processing Systems. vol. 25. Curran Associates, Inc. (2012). https://proceedings.neurips.cc/paper/2012/file/c399862d3b9d6b76c8436e924a68c45b-Paper.pdf
14. Mu, Y., Gage, F.H.: Adult hippocampal neurogenesis and its role in Alzheimer's disease. Molecul. Neurodegen. **6**(1), 85 (2011)
15. Navarro, F., et al.: Evaluating the Robustness of Self-Supervised Learning in Medical Imaging (2021)
16. Nobakht, S., Schaeffer, M., Forkert, N.D., Nestor, S., Black, S.E.P.B.: Combined Atlas and convolutional neural network-based segmentation of the hippocampus from MRI according to the ADNI harmonized protocol. Sensors **21**(7), 2427 (2021)
17. Noroozi, M., Favaro, P.: Unsupervised Learning of Visual Representations by Solving Jigsaw Puzzles. arXiv preprint arXiv:1603.09246 (2016)

18. Petersen, R., et al.: Alzheimer's Disease Neuroimaging Initiative (ADNI): clinical characterization. Neurology **74**(3), 201–209 (2010), cited By 913
19. Ronneberger, O., Fischer, P., Brox, T.: U-Net: convolutional networks for biomedical image segmentation. In: Navab, N., Hornegger, J., Wells, W.M., Frangi, A.F. (eds.) MICCAI 2015. LNCS, vol. 9351, pp. 234–241. Springer, Cham (2015). https://doi.org/10.1007/978-3-319-24574-4_28
20. Roy, A.G., Conjeti, S., Navab, N., Wachinger, C.: QuickNAT: a fully convolutional network for quick and accurate segmentation of neuroanatomy. NeuroImage **186**, 713–727 (2019)
21. Setti, S.E., Hunsberger, H.C., Reed, M.N.: Alterations in hippocampal activity and Alzheimer's disease. Transl. Issue. Psychol. Sci. **3**(4), 348–356 (2017)
22. Sled, J., Zijdenbos, A., Evans, A.: A nonparametric method for automatic correction of intensity nonuniformity in MRI data. IEEE Trans. Med. Imaging **17**(1), 87–97 (1998)
23. Smith, S.M.: Fast robust automated brain extraction. Human Brain Mapping **17**(3), 143–155 (2002)
24. Taleb, A., Lippert, C., Klein, T., Nabi, M.: Multimodal Self-Supervised Learning for Medical Image Analysis. arXiv preprint arxiv:1912.05396 (2019)

Concurrent Ischemic Lesion Age Estimation and Segmentation of CT Brain Using a Transformer-Based Network

Adam Marcus[1(✉)], Paul Bentley[1], and Daniel Rueckert[1,2]

[1] Imperial College London, London, UK
{adam.marcus11,p.bentley,d.rueckert}@imperial.ac.uk
[2] Technische Universität München, München, Germany
daniel.rueckert@tum.de

Abstract. The cornerstone of stroke care is expedient management that varies depending on the time since stroke onset. Consequently, clinical decision making is centered on accurate knowledge of timing and often requires a radiologist to interpret Computed Tomography (CT) of the brain to confirm the occurrence and age of an event. These tasks are particularly challenging due to the subtle expression of acute ischemic lesions and their dynamic nature. Automation efforts have not yet applied deep learning to estimate lesion age and treated these two tasks independently, so, have overlooked their inherent complementary relationship. To leverage this, we propose a novel end-to-end multi-task transformer-based network optimized for concurrent segmentation and age estimation of cerebral ischemic lesions. By utilizing gated positional self-attention and CT-specific data augmentation, our method can capture long-range spatial dependencies while maintaining its ability to be trained from scratch under low-data regimes commonly found in medical imaging. Further, to better combine multiple predictions, we incorporate uncertainty by utilizing quantile loss to facilitate estimating a probability density function of lesion age. The effectiveness of our model is then extensively evaluated on a clinical dataset consisting of 776 CT images from two medical centers. Experimental results demonstrate that our method obtains promising performance, with an area under the curve (AUC) of 0.933 for classifying lesion ages ≤4.5 h compared to 0.858 using a conventional approach, and outperforms task-specific state-of-the-art algorithms.

Keywords: Stroke · Computed Tomography · Lesion age estimation · Transformer network · Image segmentation

1 Introduction

Stroke is the most frequent cause of adult disability and the second commonest cause of death worldwide [20]. The vast majority of strokes are ischemic and result from the blockage of blood flow in a brain artery, often by a blood clot.

A. Abdulkadir et al. (Eds.): MLCN 2022, LNCS 13596, pp. 52–62, 2022.
https://doi.org/10.1007/978-3-031-17899-3_6

Consequently, treatment is focused on rapidly restoring blood flow before irrevocable cell death [24]. The two main approaches are: intravenous thrombolysis, chemically dissolving the blood clot; and endovascular thrombectomy, physically removing the blood clot. Notably, the efficacy of both these treatments decreases over time until their benefit is outweighed by the risk of complications. It is for this reason that current guidelines limit when specific treatments can be given. In the case of thrombolysis, to within 4.5 h of onset [7]. Therefore, accurate knowledge of timing is central to the management of stroke. However, a significant number of strokes are unwitnessed, with approximately 25% occurring during sleep. In these cases, neuroimaging can help, with previous studies showing promising results using modalities not routinely available to patients, such as magnetic resonance imaging (MRI) and perfusion computed tomography (CT) [16,26]. Ideally, the widely-available non-contrast CT (NCCT) would be used, but this task is challenging even for detection alone, as early ischemic changes are often not visible to the naked eye.

1.1 Related Work

Several studies have attempted to delineate early ischemic changes on NCCT. The majority have used image processing techniques or machine learning methods based on hand-engineered features but more recent efforts have used deep learning [6]. Qui et al. [21] proposed a random forest voxel-wise classifier using features derived from a pre trained U-Net and achieved a Dice Similarity Coefficient (DSC) of 34.7% [11]. Barros et al. [1] used a convolutional neural network (CNN) and attained a DSC of 37%. El-Hariri et al. [6] implemented a modified nnU-Net and reported DSCs of 37.7% and 34.6% compared to two experts. To the best of the authors' knowledge, the current state-of-the-art for this task is EIS-Net [11], a 3D triplet CNN that achieved a DSC of 44.8%.

In contrast, few studies have explored using NCCT to estimate the lesion age. Broocks et al. [2] used quantitative net water uptake, originally introduced by Minnerup et al. [19], to identify patients within the 4.5 h thrombolysis time window and attained an area under the receiver operator characteristic curve (AUC) of 0.91. Mair et al. [17] introduced the CT-Clock Tool, a linear model using the attenuation ratio between ischemic and normal brain, and achieved an AUC of classifying scans $\leq$4.5 h of 0.955 with median absolute errors of 0.4, 1.8, 17.2 and 32.2 h for scans acquired $\leq$3, 3–9, 9–30 and $>$30 h from stroke onset. These studies all currently require manual selection of the relevant brain regions, and as of yet, have not utilized deep-learning methods that may allow for improved performance.

Deep learning methods have shown great potential across many domains, with convolutional architectures proving highly successful in medical imaging. Here the inductive biases of CNNs, known to increase sample efficiency [5], are particularly useful due to the scarcity of medical data. However, this may be at the expense of performance, as Transformers [27] have surpassed CNNs across many computer vision tasks. By relying on flexible self-attention mechanisms, Transformer-based models can learn global semantic information beneficial to dense prediction tasks like segmentation but typically require vast amounts of training data to do so. Recently, d'Ascoli et al. [5] have attempted to address this by introducing gated positional self-attention (GPSA), a type of self-attention with a learnable gating parameter that controls the attention paid to position versus content and can combine the benefits of both architectures.

1.2 Contributions

In this work, we propose a multi-task network to simultaneously perform the segmentation of ischemic lesions and estimate their age in CT brain imaging. The main contributions are: (1) We introduce a novel end-to-end transformer-based network to solve both lesion age estimation and segmentation. To our knowledge, this is the first time a deep learning-based method has been applied to solve the challenging task of estimating lesion age. (2) We enhance the data efficiency of our approach by integrating GPSA modules into the model and using a CT-specific data augmentation strategy. (3) To further improve the performance of our model at estimating lesion age, we suggest a new method to better combine multiple predictions by incorporating uncertainty through the estimation of probability density functions. The effectiveness of the proposed method is then demonstrated by extensive experiments.

2 Method

2.1 Network

An overview of the proposed model is presented in Fig. 1. The proposed model is based on the DETR panoptic segmentation architecture [3] with modifications to improve sample efficiency, performance, and facilitate lesion age estimation. All activation functions were changed to the Gaussian error linear unit [9] (GELU) and batch normalization was replaced with group normalization [28] to accommodate smaller batch sizes. The main components of the proposed model are: 1) a CNN backbone; 2) a transformer encoder-decoder; 3) lesion, age estimation, and bounding box prediction heads; and 4) a segmentation head.

The CNN backbone encoder extracts image features of a 2D CT slice input image. It is comprised of four ResNeXt [29] blocks and produces an activation map. This activation map is then projected to a feature patch embedding and concatenated with fixed positional encodings [4]. Rather than use a 1×1 convolution as in the original DETR architecture, we use a pyramid pooling module [30] (PPM) that has empirically been shown to increase the effective receptive field by incorporating features extracted at multiple scales.

The transformer encoder-decoder learns the attention between image features and predicts output embeddings for each of the $N = 10$ object queries. Where N was determined by the maximum number of lesions visible in a given slice. We use three transformer encoder blocks and one transformer decoder block following the standard architecture [27] with a couple of exceptions. First, rather than

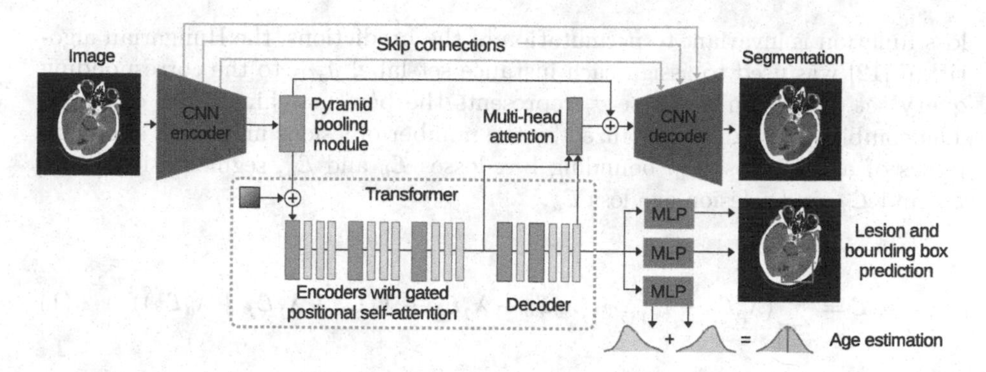

Fig. 1. Overview of the proposed model architecture.

using an auto-regressive model we decode the N objects in parallel. Second, to improve the data efficiency of the model we replace the multi-head attention layers in the encoder with GPSA layers.

The lesion, age estimation, and bounding box prediction heads are each multi-layer perceptions (MLP) and map the output embeddings of the transformer encoder-decoder to lesion, lesion age, and bounding box predictions. These heads process the queries in parallel and share parameters over all queries.

The segmentation head generates binary masks for each object instance based on attention. A two-dimensional multi-head attention layer produces attention heatmaps from the attention between the outputs of the transformer encoder and decoder. These heatmaps are then upscaled by a U-Net [23] type architecture with long skip connections between the CNN encoder and decoder blocks.

2.2 Data Augmentation

To improve the generalizability of our model and prevent overfitting due to limited training data, we adopted a CT-specific augmentation strategy with geometric and appearance transforms. Geometric transforms included: random axial plane flips; ±5% isotropic scaling; ±20 mm translation; and ±0.5rad axial otherwise ±0.1rad plane rotation. Appearance transforms included an intensity transform introduced by Zhou et al. [31] and a transform we propose to account for the slice thickness variation often present in CT datasets. Regions of the brain are area interpolated to a random slice thickness, ranging from 1–3 mm to match the sizes in our dataset, then upscaled back to their original shape.

2.3 Loss Function

We use a combined loss function to enable direct set prediction. The set prediction $\hat{y} = \{\hat{y_i} = \{\hat{p_i}, \hat{b_i}, \hat{s_i}, \hat{a_i}\}\}_{i=1}^{N}$ consists of the lesion probability $\hat{p_i} \in \mathbb{R}^2$ (lesion or no lesion), bounding box $\hat{b_i} \in \mathbb{R}^4$, segmentation mask $\hat{s_i} \in \mathbb{R}^{512 \times 512}$, and lesion age quantiles $\hat{a_i} \in \mathbb{R}^3$ for each of the N object queries. To ensure the loss function is invariant to permutation of the predictions, the Hungarian algorithm [12] was used to assign each instance set label $y_{\sigma(i)}$ to the corresponding query set prediction $\hat{y_i}$ where σ_i represents the best matching order of labels. The combined loss $\mathcal{L}$ is normalized by the number of lesions in a batch and comprises of a lesion loss $\mathcal{L}_p$, bounding box losses $\mathcal{L}_b$ and $\mathcal{L}_g$, segmentation losses $\mathcal{L}_f$ and $\mathcal{L}_d$, and a lesion age loss $\mathcal{L}_a$.

$$\mathcal{L} = \sum_{i=1}^{N}(\lambda_p\mathcal{L}_p + \mathbb{1}_{\{p_i \neq \emptyset\}}(\lambda_b\mathcal{L}_b + \lambda_g\mathcal{L}_g + \lambda_a\mathcal{L}_a + \lambda_f\mathcal{L}_f + \lambda_d\mathcal{L}_d)) \tag{1}$$

We used cross-entropy for the lesion loss $\mathcal{L}_p$. For the bounding box losses, L1 loss $\mathcal{L}_b$ and the generalized intersection over union [22] $\mathcal{L}_g$ were used. The segmentation losses comprised of Focal loss $\mathcal{L}_f$ with $\alpha = 0.25$ and $\gamma = 2$ as recommended by Lin et al. [14], and Dice loss [18] $\mathcal{L}_d$. To enable the uncertainty of lesion age estimates to be quantized, we used quantile loss for the lesion age loss $\mathcal{L}_a$. We predict three quantiles, assuming that estimates for lesion age are normally distributed, that would correspond to minus one standard deviation from the mean, the mean, and plus one standard deviation from the mean. These can be calculated using ϕ, the cumulative distribution function (CDF) of the standard normal distribution: $\tau_1 = \phi(-1) \approx 0.159$; $\tau_2 = 0.5$; $\tau_3 = \phi(1) \approx 0.841$.

$$\mathcal{L}_a(a_{\sigma(i)}, \hat{a_i}) = \sum_{j=1}^{3} \max((1-\tau_j)||a_{\sigma(i)} - \hat{a_{i,j}}||_1,\ \tau_j||a_{\sigma(i)} - \hat{a_{i,j}}||_1) \tag{2}$$

In order to account for the varying difficulties of each task common to multi-task learning procedures, we employ a random-weighted loss function where weights are drawn from the Dirichlet distribution [13].

$$\lambda_p, \lambda_b, \lambda_g, \lambda_a, \lambda_f, \lambda_d \overset{\text{i.i.d.}}{\sim} \text{Dir}(1,1,1,1,1,1) \tag{3}$$

2.4 Inference

At inference time, we combine lesion age estimates if their associated predicted segmentation masks are connected in 3D. Given a set of K lesion age quantile predictions $\hat{a} = \{\hat{a_k}\}_{k=1}^{K}, \hat{a_k} \in \mathbb{R}^3$, we estimate probability density functions (PDF) based on the split normal distribution PDF (Eq. 4) where $\mu_k = \hat{a_{k,2}}$, $\sigma_{k,1} = \hat{a_{k,2}} - \hat{a_{k,1}}$, and $\sigma_{k,2} = \hat{a_{k,3}} - \hat{a_{k,2}}$ for each instance. The maximum argument of the sum of these probability density functions is then the combined lesion age estimate, $\hat{a_\mu}$. In the rare instances where a set of predictions produces a negative $\sigma_{k,1}$ or $\sigma_{k,2}$, we resort to the mean lesion age estimate, $\bar{\mu_k}$.

$$f(x;\mu,\sigma_1,\sigma_2) = \begin{cases} A\exp\left(-\frac{(x-\mu)^2}{2\sigma_1^2}\right) & x < \mu \\ A\exp\left(-\frac{(x-\mu)^2}{2\sigma_2^2}\right) & x \geq \mu \end{cases}, \text{ where } A = \sqrt{2/\pi}(\sigma_1+\sigma_2)^{-1} \tag{4}$$

$$\hat{a_\mu} = \underset{x}{\operatorname{argmax}} \sum_{k=1}^{K} f(x; \mu_k; \sigma_{k,1}; \sigma_{k,2}) \tag{5}$$

3 Experiments

3.1 Materials

Experiments were conducted on a dataset of 776 acute stroke patients with a known time of onset collected across two clinical sites from 2013 to 2019. Ground truth segmentation masks of 79,959 slices were produced by single manual annotation from several experts. Lesion ages measured in minutes were calculated using the time from symptom onset to imaging and log-transformed to account for skewed distribution. Patients were randomly divided such that 20% were used for testing and the remainder for training and validation. Table 1 lists the characteristics of these groups. When optimizing hyperparameters, 20% of the total dataset was used for validation. Full ethical approval was granted by Wales REC 3 reference number 16/WA/0361.

Table 1. Population characteristics of the clinical dataset.

Characteristic	Train and validation set (n = 627)	Test set (n = 149)
Age (years), median (IQR)	74.9 (63.9–82.8)	74.7 (63.1–83.0)
Male sex, n (%)	317 (50.6%)	71 (47.7%)
ASPECTS, median (IQR)	9 (8–10)	9 (8–10)
NIHSS on admission, median (IQR)	13 (7–20)	13 (7–19)
Time from symptom onset to CT (minutes), median (IQR)	232 (109–1212)	253 (110–1325)

IQR = Interquartile range; ASPECTS = Alberta stroke programme early CT score; NIHSS = National Institutes of Health Stroke Scale

To evaluate lesion segmentation, we compared mean DSC and intersection over union (IOU) between model predictions and expert segmentation's on a per-subject level. The Mann-Whitney U test was used to determine significance. For lesion age, we excluded subjects with lesions of different ages and calculated the coefficient of determination (R^2), mean absolute error (MAE), and root mean squared error (RMSE). We also evaluated the classification of lesion age within 4.5 h of onset using accuracy (ACC) and AUC.

All models were implemented using PyTorch version 1.10 and trained from scratch for 100 epochs on a computer with 3.80GHz Intel® Core™ i7-10700K CPU and an NVIDIA GeForce RTX 3080 10GB GPU. The AdamW [15] optimizer was used with a weight decay of 10^{-4}. Learning rate was adjusted from

10^{-6} to 10^{-2} per-epoch using a cyclical schedule [25] and exponentially decayed per-cycle with $\gamma = 0.92$. Gradient clipping was applied to ensure a maximal gradient norm of 0.1. We also employed the stochastic weight averaging [10] for the last 5 cycles. During training, lesion containing regions were linearly sampled from the original volumes to a uniform size, $512 \times 512 \times 1$ for 2D and $128 \times 128 \times 48$ for 3D models, with a spatial resolution of $0.45 \times 0.45 \times 0.8\,\text{mm}^3$. Pixel intensities were clipped based on the 0.5 and 99.5th percentile then normalized using Z-score. Inference of the models required about 14 s per subject.

3.2 Results

Comparison with Baseline. We first compare our proposed model to task-specific deep-learning algorithms due to the absence of established methods to jointly perform segmentation and regression. The quantitative results are shown in Table 2. For segmentation, we compare against 2D and 3D U-Net [23] using the same Focal and Dice loss function. In this task, our proposed model performs slightly better, with significant (p value ≤ 0.05) increases in DSC and IOU at the expense of greater computational demands. Notably, despite the proposed model being 2D in nature, it performed competitively against 3D U-Net, suggesting that for lesion segmentation the ability to capture global semantic information may outweigh the benefits of learning volumetric relations. These findings are also supported by qualitative evaluation as seen in Fig. 2. For lesion age estimation, we first trained a linear model based on intensity using a similar methodology to Mair et al. [17]. We also trained ResNet-50 [8] and ResNeXt-50-32 × 4d [29] models using the same quantile loss function. Compared against these models, our proposed method outperforms them by large margins for all metrics tested. It seems, therefore, that explicit supervised learning of both tasks may be mutually beneficial and is particularly useful in estimating lesion age.

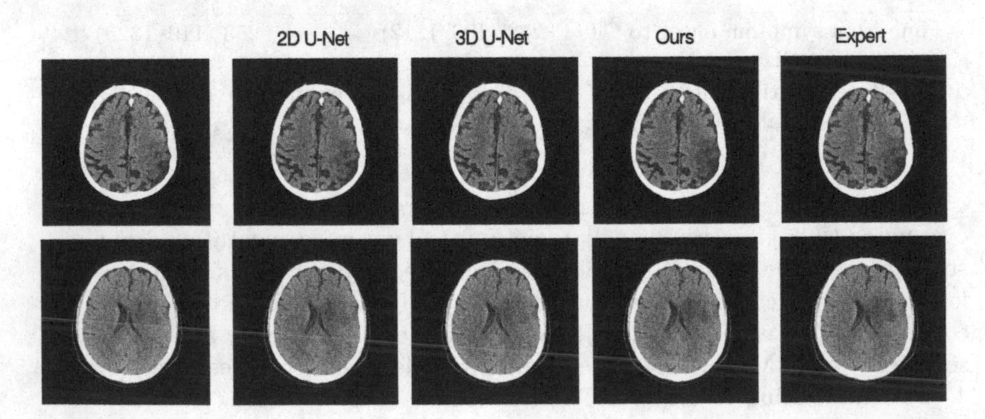

Fig. 2. Example lesion segmentations of our method compared to the baseline models.

Table 2. Lesion age estimation and segmentation (mean ± standard deviation) results obtained by our method and ablation variants compared to the baseline models.

		Regression			Classification		Segmentation	
Model	Size	R^2	MAE	RMSE	AUC	ACC	DSC	IOU
Intensity GLM	2	0.365	0.816	1.021	0.858	79.5	—	—
ResNet-50	24M	0.308	0.862	1.115	0.906	83.1	—	—
ResNeXt-50	23M	0.402	0.800	1.037	0.908	86.5	—	—
Ours	40M	**0.513**	0.680	**0.935**	**0.933**	**88.5**	38.2 ± 24.2	**26.6 ± 21.0**
2D U-Net	8M	—	—	—	—	—	35.3 ± 30.0	26.2 ± 26.2
3D U-Net	39M	—	—	—	—	—	36.7 ± 28.2	26.4 ± 26.4
ResNet-50 (L_1)	24M	0.297	0.866	1.124	0.904	81.7	—	—
ResNeXt-50 (L_1)	23M	0.396	0.809	1.112	0.907	84.5	—	—
Ours (L_1)	40M	0.503	**0.636**	0.944	0.912	86.5	**38.2 ± 24.1**	26.3 ± 21.1
Ours (no PPM)	30M	0.330	0.733	1.097	0.874	79.7	36.0 ± 24.0	24.9 ± 20.8
Ours (no GPSA)	40M	0.449	0.664	0.995	0.913	83.8	35.4 ± 24.6	24.9 ± 20.7
Ours (no RLW)	40M	0.402	0.675	1.036	0.904	83.4	35.0 ± 25.0	24.5 ± 21.5
Ours (no DA)	40M	0.025	0.945	1.357	0.756	71.6	31.6 ± 24.6	21.7 ± 20.6

MAE = Mean absolute error; AUC = Area under the receiver operator characteristic curve; ACC = Accuracy; GLM = Generalized linear model; L_1 = L1 loss; PPM = Pyramid pooling module; GPSA = Gated positional self-attention; RLW = Random loss weighting; DA = Data augmentation

Comparison with the State-of-the-Art. There are few works that we can compare our results. For segmentation, we are aware of only two studies [1,6] that used ground truth NCCT annotations. As argued by El-Hariri et al. [6], direct comparison with studies using annotations from other modalities such as MRI is hindered by the different underlying physiological processes which lead to visible changes. Compared with these studies, the proposed model performs slightly better on this challenging task with a DSC of 38.2% compared to 37% by Barros et al. [1] and 37.7% by El-Hairi et al. [6]. For lesion age estimation, the proposed model achieved an AUC of 0.933 for classifying whether a stroke event is within 4.5 h of onset. Similar to the predominately manual methods by Broocks et al. [2] and Mair et al. [17] with reported AUC of 0.91 and 0.955, respectively. However, we note that due to the dynamic nature of ischemia, the classification of older lesions is considerably easier. Therefore, the difficulty of this task is highly dependent on the distribution of lesion ages in the dataset, and without an open benchmark, objective assessment against other methods is limited. This is further supported by our intensity model achieving an AUC of only 0.858 using a similar methodology to these studies.

Ablation Study. We conducted a series of experiments, shown in Table 2, to verify the effectiveness of our method and justify its design decisions. First, we observe that our data augmentation strategy appears to have the largest impact on lesion age estimation and segmentation performance. Second, using GPSA, PPM, and RLW rather than equally weighted losses provide benefits primarily to age estimation with comparatively little effect on segmentation. Finally, we note a consistent increase in lesion age estimation performance gained by using our proposed quantile loss based method across all tested models.

4 Conclusion

In this paper, we proposed a novel transformer-based network for concurrent ischemic lesion segmentation and age estimation of CT brain. By incorporating GPSA layers and using a modality-specific data augmentation strategy, we enhanced the data efficiency of our method. Furthermore, we improved lesion age estimation performance by better combining multiple predictions through the incorporation of uncertainty. Extensive experiments on a clinical dataset demonstrated the effectiveness of our method compared to conventional and task-specific algorithms. Future work includes further prospective clinical validation and exploring the extension of the model to 3D.

Acknowledgements. Adam Marcus is supported by the UKRI CDT in Artificial Intelligence for Healthcare http://ai4health.io (Grant No. P/S023283/1).

References

1. Barros, R.S., et al.: Automated segmentation of subarachnoid hemorrhages with convolutional neural networks. Inform. Med. Unlocked **19**, 100321 (2020)
2. Broocks, G., et al.: Lesion age imaging in acute stroke: water uptake in CT versus DWI-flair mismatch. Ann. Neurol. **88**(6), 1144–1152 (2020)
3. Carion, N., Massa, F., Synnaeve, G., Usunier, N., Kirillov, A., Zagoruyko, S.: End-to-end object detection with transformers. In: Vedaldi, A., Bischof, H., Brox, T., Frahm, J.-M. (eds.) ECCV 2020. LNCS, vol. 12346, pp. 213–229. Springer, Cham (2020). https://doi.org/10.1007/978-3-030-58452-8_13
4. Cordonnier, J.B., Loukas, A., Jaggi, M.: On the relationship between self-attention and convolutional layers. arXiv preprint arXiv:1911.03584 (2019)
5. d'Ascoli, S., Touvron, H., Leavitt, M.L., Morcos, A.S., Biroli, G., Sagun, L.: Convit: Improving vision transformers with soft convolutional inductive biases. In: International Conference on Machine Learning. pp. 2286–2296. PMLR (2021)
6. El-Hariri, H., et al.: Evaluating NNU-Net for early ischemic change segmentation on non-contrast computed tomography in patients with acute ischemic stroke. Computers in biology and medicine p. 105033 (2021)
7. Hacke, W., et al.: Thrombolysis with alteplase 3 to 4.5 hours after acute ischemic stroke. New England J. Med. **359**(13), 1317–1329 (2008)
8. He, K., Zhang, X., Ren, S., Sun, J.: Deep residual learning for image recognition. In: Proceedings of the IEEE Conference on Computer Vision and Pattern Recognition. pp. 770–778 (2016)
9. Hendrycks, D., Gimpel, K.: Gaussian error linear units (gelus). arXiv preprint arXiv:1606.08415 (2016)

10. Izmailov, P., Podoprikhin, D., Garipov, T., Vetrov, D., Wilson, A.G.: Averaging weights leads to wider optima and better generalization. arXiv preprint arXiv:1803.05407 (2018)
11. Kuang, H., Menon, B.K., Sohn, S.I., Qiu, W.: EIS-Net: Segmenting early infarct and scoring aspects simultaneously on non-contrast CT of patients with acute ischemic stroke. Med. Image Anal. **70**, 101984 (2021)
12. Kuhn, H.W.: The Hungarian method for the assignment problem. Naval Res. Logistics Q. **2**(1–2), 83–97 (1955)
13. Lin, B., Ye, F., Zhang, Y.: A closer look at loss weighting in multi-task learning. arXiv preprint arXiv:2111.10603 (2021)
14. Lin, T.Y., Goyal, P., Girshick, R., He, K., Dollár, P.: Focal loss for dense object detection. In: Proceedings of the IEEE International Conference on Computer Vision. pp. 2980–2988 (2017)
15. Loshchilov, I., Hutter, F.: Decoupled weight decay regularization. arXiv preprint arXiv:1711.05101 (2017)
16. Ma, H., et al.: Thrombolysis guided by perfusion imaging up to 9 hours after onset of stroke. N. Engl. J. Med. **380**(19), 1795–1803 (2019)
17. Mair, G., Alzahrani, A., Lindley, R.I., Sandercock, P.A., Wardlaw, J.M.: Feasibility and diagnostic accuracy of using brain attenuation changes on CT to estimate time of ischemic stroke onset. Neuroradiology **63**(6), 869–878 (2021)
18. Milletari, F., Navab, N., Ahmadi, S.A.: V-Net: Fully convolutional neural networks for volumetric medical image segmentation. In: 2016 Fourth International Conference on 3D Vision (3DV), pp. 565–571. IEEE (2016)
19. Minnerup, J.B., et al.: Computed tomography-based quantification of lesion water uptake identifies patients within 4.5 hours of stroke onset: A multicenter observational study. Annals Neurol. **80**(6), 924–934 (2016)
20. Organization, W.H.: Global health estimates (2018). https://www.who.int/healthinfo/global_burden_disease/en/
21. Qiu, W., et al.: Machine learning for detecting early infarction in acute stroke with non-contrast-enhanced CT. Radiology **294**(3), 638–644 (2020)
22. Rezatofighi, H., Tsoi, N., Gwak, J., Sadeghian, A., Reid, I., Savarese, S.: Generalized intersection over union: A metric and a loss for bounding box regression. In: Proceedings of the IEEE/CVF Conference on Computer Vision and Pattern Recognition, pp. 658–666 (2019)
23. Ronneberger, O., Fischer, P., Brox, T.: U-Net: convolutional networks for biomedical image segmentation. In: Navab, N., Hornegger, J., Wells, W.M., Frangi, A.F. (eds.) MICCAI 2015. LNCS, vol. 9351, pp. 234–241. Springer, Cham (2015). https://doi.org/10.1007/978-3-319-24574-4_28
24. Saver, J.L.: Time is brain-quantified. Stroke **37**(1), 263–266 (2006)
25. Smith, L.N.: Cyclical learning rates for training neural networks. In: 2017 IEEE Winter Conference on Applications of Computer Vision (WACV), pp. 464–472. IEEE (2017)
26. Thomalla, G., et al.: MRI-guided thrombolysis for stroke with unknown time of onset. N. Engl. J. Med. **379**(7), 611–622 (2018)
27. Vaswani, A., et al.: Attention is all you need. Adv. Neural Inform. Proc. Syst. **30** (2017)
28. Wu, Y., He, K.: Group normalization. In: Proceedings of the European Conference on Computer Vision (ECCV), pp. 3–19 (2018)
29. Xie, S., Girshick, R., Dollár, P., Tu, Z., He, K.: Aggregated residual transformations for deep neural networks. In: Proceedings of the IEEE Conference on Computer Vision and Pattern Recognition, pp. 1492–1500 (2017)

30. Zhao, H., Shi, J., Qi, X., Wang, X., Jia, J.: Pyramid scene parsing network. In: Proceedings of the IEEE Conference on Computer Vision and Pattern Recognition, pp. 2881–2890 (2017)
31. Zhou, Z., et al.: Models genesis: generic autodidactic models for 3D medical image analysis. In: Shen, D. (ed.) MICCAI 2019. LNCS, vol. 11767, pp. 384–393. Springer, Cham (2019). https://doi.org/10.1007/978-3-030-32251-9_42

Weakly Supervised Intracranial Hemorrhage Segmentation Using Hierarchical Combination of Attention Maps from a Swin Transformer

Amirhossein Rasoulian[1(✉)], Soorena Salari[1], and Yiming Xiao[1,2,3]

[1] Department of Computer Science and Software Engineering, Concordia University, Montreal, Canada
ah.rasoulian@gmail.com
[2] PERFORM Centre, Concordia University, Montreal, Canada
[3] Applied AI Institute, Concordia University, Montreal, Canada

Abstract. Intracranial hemorrhage (ICH) is a potentially life-threatening emergency due to various causes. Rapid and accurate diagnosis of ICH is critical to deliver timely treatments and improve patients' survival rates. Although deep learning techniques have become the state-of-the-art in medical image processing and analysis, large training datasets with high-quality annotations that are expensive to acquire are often necessary for supervised learning. This is especially true for image segmentation tasks. To facilitate ICH treatment decisions and tackle this issue, we proposed a novel weakly supervised ICH segmentation method utilizing a hierarchical combination of self-attention maps obtained from a Swin transformer, which is trained through an ICH classification task with categorical labels. We developed and validated the proposed technique using two public clinical CT datasets (RSNA 2019 Brain CT hemorrhage & PhysioNet). As an exploratory study, we compared two different learning strategies (binary classification vs. full ICH subtyping) to investigate their impacts on self-attention and our weakly-supervised ICH segmentation method. As the first to perform ICH detection and weakly supervised segmentation with a Swin transformer, our algorithm achieved a Dice score of 0.407 ± 0.225 for ICH segmentation while delivering high accuracy in ICH detection (AUC = 0.974).

Keywords: Weak supervision · Swin transformer · Attention · Intracranial hemorrhage · Segmentation · Computerized tomography

1 Introduction

Intracranial Hemorrhage (ICH) is the most deadly type of cerebrovascular disease, accounting for 10–15% of all stroke cases [1,14]. The outcome is highly correlated with the hemorrhage volume, which is susceptible to enlarge in the

A. Abdulkadir et al. (Eds.): MLCN 2022, LNCS 13596, pp. 63–72, 2022.
https://doi.org/10.1007/978-3-031-17899-3_7

first three hours [13]. Thus, there is a high risk for ICH to turn into a secondary brain injury or even death if it is not treated in time. Depending on the location of hemorrhage in the brain, ICH can be divided into five subtypes: Intraventricular (IVH), Intraparenchymal (IPH), Subarachnoid (SAH), Epidural (EDH), and Subdural (SDH). Treatment methods must be tailored towards specific ICH subtypes, and a surgery is done only if the location of hemorrhage is favorable. Rapid and accurate detection and quantification of ICH is therefore crucial in choosing correct treatments and thus reduction of patient mortality. With quick imaging time and good accessibility, computerized tomography (CT) is commonly used in the clinic to assess ICH.

Previous developments in convolutional neural networks (CNNs) have resulted in a great number of fast and accurate solutions in computer-assisted diagnosis and treatment decisions, in the forms of image classification and/or segmentation, including those for the care of ICH [9]. One issue with the CNNs is their limited capacity to encode long-range spatial information, but it may affect ICH detection/subtyping accuracy as the location of hemorrhage is directly relevant to the diagnosis. Recently, Dosovitskiy et al. [6] introduced the Vision Transformer (ViT), which has attracted significant interests for vision tasks, especially in the context of medical imaging [4,5], where multi-head attention mechanisms are used to encode the contextual relationship between image patches (as tokens). However, compared with CNNs, the ViT has low locality inductive biases (e.g., translational invariant features). As a recent variant to mitigate the drawback of the ViT, the Swin transformer [11] is an efficient hierarchical transformer that gradually reduces the number of tokens by merging image patches and computing attentions in non-overlapping local windows. To the best of our knowledge, the Swin transformer has not been used for ICH detection or segmentation. For CNNs and especially transformer-based models, a large amount of training data is necessary. However, annotating medical images is laborious and time-consuming, especially for segmentation tasks. One way to mitigate this problem is through weak supervision [22], where more accessible or coarse annotations (e.g., categorical labels or bounding boxes) are used to generate more refined ones, such as segmentation masks.

In this work, we built a novel weakly supervised framework for ICH segmentation leveraging the attention maps generated from a Swin transformer, which is trained using categorical labels for ICH detection based on public databases. As an exploratory investigation, our study has three major contributions. **First**, the Swin transformer is used for ICH detection for the first time. **Second**, we proposed a new method to obtain ICH segmentation by leveraging the hierarchical combination of self-attention maps from the trained ICH detection transformer, and demonstrated its feasibility and performance. **Lastly**, to examine the impact of learning tasks on self-attention maps and weakly supervised segmentation, we compared the segmentation performance for two Swin transformers based on (1) binary classification (presence of hemorrhage or not) and (2) multi-label classification (detailed ICH subtypes and with/without ICH).

2 Related Works

Several techniques have been proposed for the detection and segmentation of ICH. An excellent recent review is provided by Hssayeni et al. [9], with almost all using supervised learning strategies in semi-automatic and automatic manners, achieving the area-under-the-curve (AUC) of 0.846∼0.975 for binary classification (ICH vs. without ICH) and 0.93∼0.96 for ICH subtyping. For deep learning-based approaches, fully convolutional networks (FCNs) [3] and recurrent neural networks (RNNs) [21] models were often used, and the accuracy in ICH vs. without ICH classification was shown to be higher than ICH sub-typing in general [9]. With interests in explainable CNNs, attention mechanisms have been deployed to enhance detection accuracy and visualize the region of interest for the classification results [15]. The latter also inspired their application for weakly supervised brain lesion/hemorrhage segmentation, which has been attempted by only a few [12,19]. Earlier, Wu et al. [19] employed refined 3D Class-Activation Maps (CAMs) to learn a representation model for brain lesion segmentation and achieved a 0.3827 mean Dice score on the Ischemic Stroke Lesion Segmentation (ISLES) dataset (multi-spectral MRI). Similarly, Nemcek et al. [12] detected the location of ICH as bounding boxes in axial brain CT slices using the local extrema of attention maps obtained from a ResNet-like binary classification CNN, and they achieved a mean Dice of 0.58 for the lesion bounding boxes. So far, self-attention has not been experimented for weakly supervised ICH segmentation, thus motivating us to explore it in this study.

3 Proposed Methods

An overview of the proposed weakly supervised segmentation technique is depicted in Fig. 1, where it is divided into three components. First, we train a deep learning (DL) model with a Swin transformer as the backbone for categorical classification of ICH. Then, during test time, we obtain hierarchical layer-wise attention maps for the input image from the trained model. Finally, segmentation is achieved by binarizing the hemorrhage localization map made by combining the window attention maps and soft tissue intensity information. Note that one patient may have multiple ICH subtypes. Since the CT data were from several clinical centers with different slice thicknesses, we decided to implement our algorithm for 2D axial CT slices.

ICH Classification: For the proposed technique, we used the Swin-B transformer pretrained and finetuned on ImageNet1K and ImageNet21K datasets [18]. Each two successive Swin transformer blocks have window multi-head self-attention (W-MSA) and shifted window multi-head self-attention (SW-MSA) units for computing attention weights (see Fig. 2a) [11]. Here, the shifted windowing scheme helps establish connections between windows, in comparison to the ViT. To investigate the impact of different arrangments of categorical learning on the self-attention maps and thus the proposed weakly supervised segmentation, we trained two versions of the Swin transformer model for 1) binary

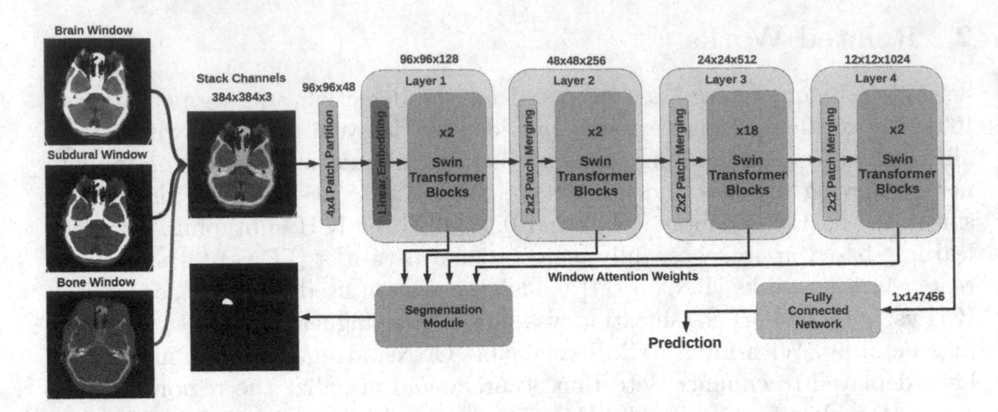

Fig. 1. An overview of the proposed weakly supervised ICH segmentation algorithm

classification (ICH vs. without ICH) and 2) multi-label classification (recognizing 5 ICH subtypes and with/without ICH). To address the issue of the imbalanced dataset, we use the focal binary cross-entropy [10] loss function to train our model:

$$loss = \frac{1}{N} \sum_{k=1}^{N} Y_k \cdot (1 - y_k)^{\gamma} \cdot log(y_k) + (1 - Y_k) \cdot y_k^{\gamma} \cdot log(1 - y_k) \tag{1}$$

Here N, Y, y, and γ are batch size, ground-truth label, sigmoid of predicted output, and the focal loss focusing parameter, respectively. As Lin et al. [10] suggested, we set the value of $\gamma = 2$. For multi-label classification, the overall loss is the weighted average of subtypes' losses computed above, where each of five subtypes' weight is 1, and ICH vs. without ICH weight is 2.

Attention Map Generation: For our technique, we decided to employ the raw attention weights of all layers to obtain the relevant attention maps for weakly supervised segmentation, instead of the more commonly used visualization of class activation mapping (CAM). This is due to two reasons. First, we would like to fully leverage the relevant information from earlier layers considering the Swin transformer architecture. Second, without gradient computation, the processing can be more efficient.

In previous attempts to visualize attention weights with the ViT, an additional classification token was added to the image patches, and after multiplying the attention weights of all layers, only this token's weight was retrieved as the attention map [2,6]. However, as the Swin transformer uses a windowing method, adding another token corrupts the window division. Besides, attention weights at every two successive MSA units correspond to regular and shifted image patches, and multiplying them is meaningless. Hence, instead of multiplying weights, we compute the attention map at each unit, and then we multiply their respective maps. Here, Fig. 2 shows the steps of producing the layer-wise

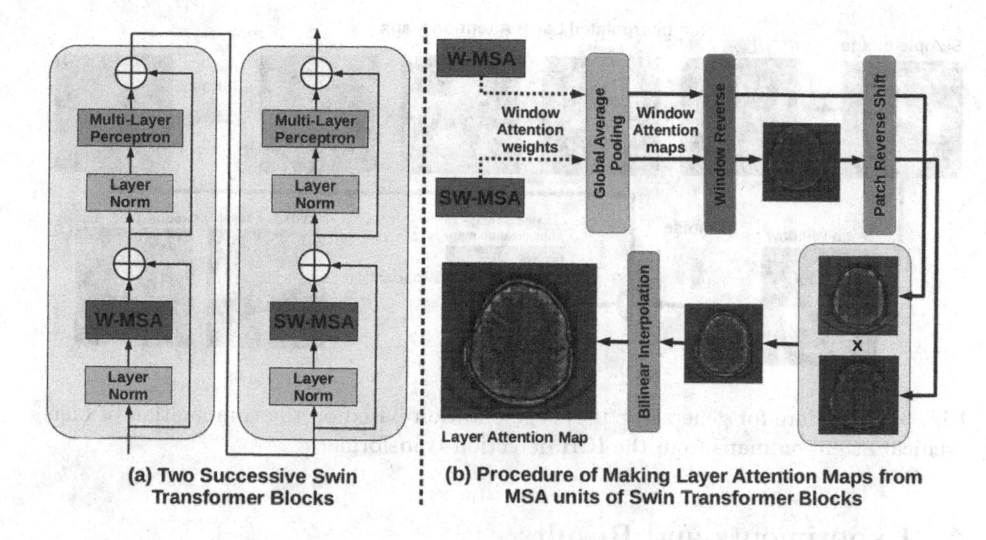

Fig. 2. Generation of layer-wise attention maps from the Swin transformer

attention map from two successive Swin transformer blocks. First, we perform Global Average Pooling (GAP) on all tokens' attention weights; Then, a full-image map is reproduced by concatenating window-wise maps. Note that an additional step of reverse shifting is needed for SW-MSA units, and the results from W-MSA and SW-MSA are multiplied to produce the layer attention map. Inspired by [8,17,20], we finally combine different layers' attention maps at the resolution of the input image, and bilinear interpolation is used when matching the resolutions across the maps at different layers/hierarchies. This technique helps produce more precise attention visualization for segmentation purposes. Lastly, the hemorrhage localization map (see Fig. 3) is produced by multiplying the resulting self-attention map with the "brain-tissue window" channel from the CT slice to enhance the discriminative power for ICH identification.

Discrete Segmentation: The final discrete ICH segmentation is obtained by binarizing the hemorrhage localization map (see Fig. 3). We experimented with three binarization techniques, including simple thresholding, Otsu's method, and k-means, and selected simple thresholding as the optimal choice due to its performance. For simple thresholding, the threshold value is computed as $t = S \times M_{max}$ where M_{max} is the maximum intensity in the hemorrhage localization map, and S is a scalar. We used 10-fold cross-validation on the test data to find an appropriate value for S between 0 to 1 with a step size of 0.01.

To compare with the proposed technique, we also implemented a similar weakly supervised ICH segmentation method based on binary ICH classification with the Swin transformer and GradCAM [16], which was implemented in two verions: one only on the last layer and the other with a similar hierarchical approach to obtain the attention maps.

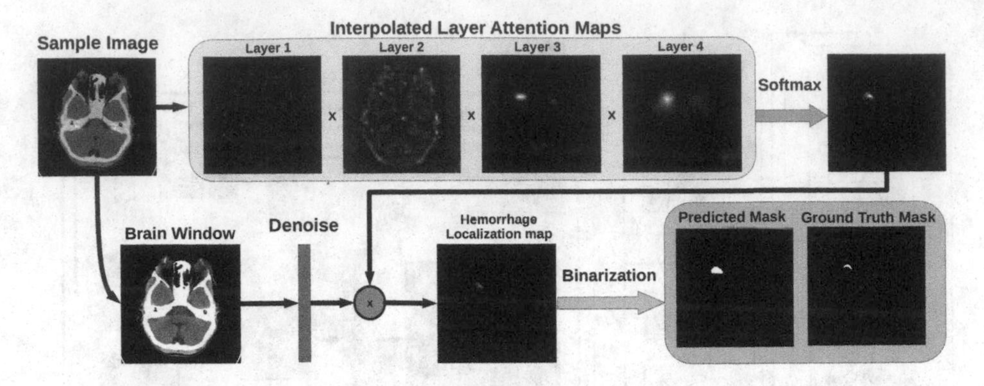

Fig. 3. Procedure for generating ICH segmentation based on the combination of hierarchical attention maps from the ICH detection transformer

4 Experiments and Results

4.1 Dataset

To implement and validate our proposed algorithm, we employed the public RSNA 2019 ICH [7] and PhysioNet [9] CT datasets. The RNSA dataset contains 752,803 CT slices, with each slice annotated with ICH subtypes. On the other hand, the PhysioNet dataset has 2,814 CT slices, and ICH was manually segmented while ICH subtypes are also provided. Only the RSNA dataset was used for training, and we used the PhysioNet dataset as a separate testing set to examine the performance of our model. For each CT slice, brain, subdural and bone windows created using the suggested parameters provided in the relevant data publications [7,9] were stacked to create a three-channel image and are downsampled to 384×384 pixels.

4.2 Imlpementation and Evaluation

The training dataset (RSNA2019 ICH) was randomly split into 90% and 10% for the training and validation sets. We employed the AdamW optimizer with an initial learning rate of 1e-5 for model training. In addition, an early stopping with patience = 3 was used to stop training if the validation loss did not decrease for three consecutive epochs. To improve the robustness of our model, data augmentation techniques including random left-right flipping, image rotation, and Gaussian noise addition were also used. The focal binary cross-entropy was used as the loss function to tackle the imbalanced data in the training dataset, where much more CT slices without ICH exist. The network was trained on a desktop computer with an Intel Core i9 CPU and a NVIDIA GeForce RTX 3090 GPU with 24 GB memory. To test the performance of the ICH detection accuracy and the performance of hemorrhage segmentation, the PhysioNet-ICH data was used. The accuracy, AUC, specificity, and F1-score were evaluated for the classification

Table 1. ICH detection and weakly supervised segmentation results for binary and multi-label ICH classification models (reported values for Dice are mean±std)

	ICH detection				ICH segmentation - Dice		
	Accuracy	AUC	Specificity	F1-score	Simple thresholding	Otsu's method	k-means
Binary	0.953	0.974	0.973	0.791	0.407 ± 0.225	0.383 ± 0.228	0.326 ± 0.228
Multi-label	0.952	0.975	0.979	0.776	0.324 ± 0.237	0.316 ± 0.246	0.268 ± 0.229

tasks, and for segmentation, the Dice coefficient is reported. When assessing the multi-label classification model against the binary classification one, an image is categorized as ICH if any subtypes are detected. Thus, the differences in ICH detection between the two DL models were confirmed using a chi-square test, and the Dice coefficients for segmentation performance between the two models were compared using a two-sided paired-sample t-test.

4.3 Results

The results of our experiments are listed in Table 1 for the automatic detection and weakly supervised segmentation of ICH when employing binary and multi-label classification tasks. In terms of the quality of ICH detection, there is no significant difference between the two proposed DL models for ICH vs. without ICH classification ($p > 0.05$), while the binary classification achieves an AUC of 0.974. Regarding ICH subtyping, we have achieved the AUCs of 0.941, 0.976, 0.996, 0.965, and 0.984 for EDH, IPH, IVH, SAH, and SDH, respectively. For hemorrhage segmentation, the binary classification model yielded a mean Dice of 0.407 (with simple thresholding), which is significantly higher than the multi-label counterpart ($p < 0.05$). The same trend holds for the other two image binarization methods ($p < 0.05$). When comparing different binarization methods, simple thresholding offers the best results, potentially due to the high imbalance between ICH and non-ICH pixels. Furthermore, as an ablation study, the GradCAM-based methods, when applied to the final layer and with a similar hierarchical approach, could achieve 0.187 and 0.100 mean Dice scores (also using the simple thresholding method), respectively, which are far worse than our proposed method. Finally, a qualitative demonstration of the segmentation results is shown in Fig. 4 for five different cases (each case per column) between the proposed method and the GradCAM approach. With the visual demonstrations, we can see that the final segmentation with the proposed method produces better results than the GradCAM approach, which provides good coverage for the hemorrhage regions, but often over-estimates the extent.

5 Discussion

As the first attempt to use the Swin transformer for ICH detection, we obtained an accuracy of $AUC = 0.974$. For both cases of binary and multi-label classifications, these results are in line with or better than previous reports [9]. Since there

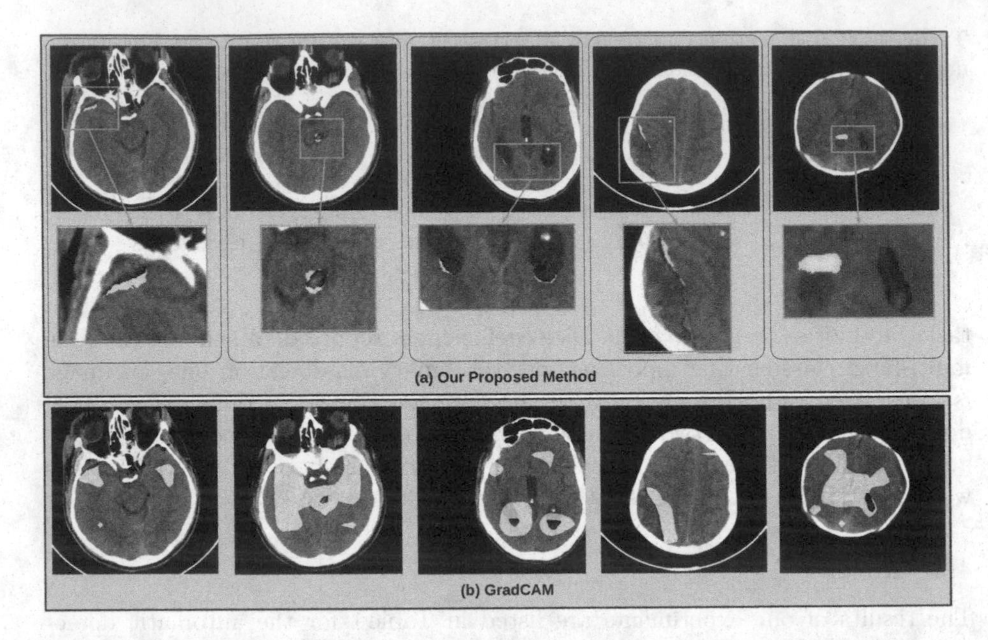

Fig. 4. Demonstration of weakly supervised ICH segmentation with five cases with close-up views. Green = True Positive, Blue = False Negative, Red = False Positive. (Color figure online)

were very few studies on weakly-supervised pixel/voxel-wise ICH segmentation, it is difficult to assess our method against the state-of-the-art. The closest prior work is the technique by [19], which was employed to segment stroke lesions from multi-spectral MRI with a Dice score of 0.3827. For ICH segmentation, a similar but potentially more challenging task due to the small size, irregular shape, and subtle contrast of the target in CT, our technique has achieved a higher Dice score (0.407). As an additional reference, in the original data paper of PhysioNet [9], a supervised U-Net achieved a Dice of 0.315 for ICH segmentation. Although the overall performance of weakly supervised brain lesion/hemorrhage segmentation is still inferior to the supervised counterparts, the relevant explorations, like the presented study are valuable in mitigating the heavy reliance on detailed image annotations.

The attention mechanism has been popular to improve the transparency of deep learning algorithms and has been the focus of many weakly supervised segmentation algorithms [12,19]. Nevertheless, the impact of different learning strategies on attention maps was rarely investigated. In this exploratory study, we examined such an impact through the example of ICH detection by comparing binary and multi-label classifications. As our weakly-supervised ICH segmentation heavily relies on the resulting self-attention maps, the segmentation accuracy also reflects how well the network focuses on the relevant regions for the designated diagnostic task. Based on our observations using the Swin transformer, binary classification offers better overlap between the network's attention and the relevant pathological region, while both strategies offer similar performance for ICH detection (grouping all subtypes as ICH). Future exploration is still needed to better understand the observed trend.

Several aspects can still be explored to further improve our segmentation performance in the future. First, instead of 2D slice processing, inter-slice or 3D spatial information may be incorporated to enhance the performance of ICH detection and segmentation. Second, more efficient and elaborate learning-based methods can be devised to further refine the initial segmentation obtained with self-attention maps to allow better segmentation accuracy.

6 Conclusion

In conclusion, leveraging the Swin transformer and public datasets, we have developed a framework for weakly supervised segmentation of ICH based on categorical labels. To tackle the issue of limited and expensive training data for ICH segmentation, we have showcased the feasibility of this approach and further demonstrated the benefit of binary classification over multi-label classification in weakly supervised segmentation. With these insights, future studies could further improve the proposed technique's accuracy and robustness.

Acknowledgement. This work was supported by a Fonds de recherche du Québec - Nature et technologies (FRQNT) Team Research Project Grant (2022-PR296459).

References

1. Apostolaki-Hansson, T., Ullberg, T., Pihlsgård, M., Norrving, B., Petersson, J.: Prognosis of intracerebral hemorrhage related to antithrombotic use: an observational study from the Swedish stroke register (riksstroke). Stroke **52**(3), 966–974 (2021)
2. Chefer, H., Gur, S., Wolf, L.: Transformer interpretability beyond attention visualization. In: Proceedings of the IEEE/CVF Conference on Computer Vision and Pattern Recognition, pp. 782–791 (2021)
3. Cho, J., et al.: Improving sensitivity on identification and delineation of intracranial hemorrhage lesion using cascaded deep learning models. J. Digit. Imaging **32** 450–461 (2018)
4. Dai, Y., Gao, Y., Liu, F.: Transmed: transformers advance multi-modal medical image classification. Diagnostics **11**(8), 1384 (2021)
5. Dalmaz, O., Yurt, M., Çukur, T.: Resvit: Residual vision transformers for multimodal medical image synthesis. IEEE Trans. Med. Imaging, 1 (2022)
6. Dosovitskiy, A., et al.: An image is worth 16x16 words: Transformers for image recognition at scale. In: International Conference on Learning Representations (2021)
7. Flanders, A.E., et al.: Construction of a machine learning dataset through collaboration: the RSNA 2019 brain CT hemorrhage challenge. Radiol.: Artif. Intell. **2**(3) (2020)

8. Gu, Y., Yang, K., Fu, S., Chen, S., Li, X., Marsic, I.: Multimodal affective analysis using hierarchical attention strategy with word-level alignment. In: Proceedings of the Conference Association for Computational Linguistics Meeting vol. 2018, p. 2225. NIH Public Access (2018)
9. Hssayeni, M.D., Croock, M.S., Salman, A.D., Al-khafaji, H.F., Yahya, Z.A., Ghoraani, B.: Intracranial hemorrhage segmentation using a deep convolutional model. Data **5**(1), 14 (2020)
10. Lin, T.Y., Goyal, P., Girshick, R., He, K., Dollár, P.: Focal loss for dense object detection. In: Proceedings of the IEEE International Conference on Computer Vision, pp. 2980–2988 (2017)
11. Liu, Z., et al.: Swin transformer: Hierarchical vision transformer using shifted windows. In: Proceedings of the IEEE/CVF International Conference on Computer Vision, pp. 10012–10022 (2021)
12. Nemcek, J., Vicar, T., Jakubicek, R.: Weakly supervised deep learning-based intracranial hemorrhage localization. arXiv preprint arXiv:2105.00781 (2021)
13. Qureshi, A., Palesch, Y.: Antihypertensive treatment of acute cerebral hemorrhage (ATACH) ii: design, methods, and rationale. Neurocrit. Care **15**(3), 559–576 (2011)
14. Rajashekar, D., Liang, J.W.: Intracerebral hemorrhage. In: StatPearls [Internet]. StatPearls Publishing (2021)
15. Salehinejad, H., et al.: A real-world demonstration of machine learning generalizability in the detection of intracranial hemorrhage on head computerized tomography. Scientific Reports 11(17051) (2021)
16. Selvaraju, R.R., Cogswell, M., Das, A., Vedantam, R., Parikh, D., Batra, D.: Grad-cam: Visual explanations from deep networks via gradient-based localization. In: Proceedings of the IEEE International Conference on Computer Vision, pp. 618–626 (2017)
17. Sindagi, V.A., Patel, V.M.: Ha-CNN: Hierarchical attention-based crowd counting network. IEEE Trans. Image Process. **29**, 323–335 (2019)
18. Wightman, R.: Pytorch image models. https://github.com/rwightman/pytorch-image-models (2019)
19. Wu, K., Du, B., Luo, M., Wen, H., Shen, Y., Feng, J.: Weakly supervised brain lesion segmentation via attentional representation learning. In: Shen, D. (ed.) MICCAI 2019. LNCS, vol. 11766, pp. 211–219. Springer, Cham (2019). https://doi.org/10.1007/978-3-030-32248-9_24
20. Yang, Z., Yang, D., Dyer, C., He, X., Smola, A., Hovy, E.: Hierarchical attention networks for document classification. In: Proceedings of the 2016 Conference of the North American Chapter of the Association for Computational Linguistics: Human Language Technologies, pp. 1480–1489 (2016)
21. Ye, H., et al.: Precise diagnosis of intracranial hemorrhage and subtypes using a three-dimensional joint convolutional and recurrent neural network. Eur. Radiol. **29**(11), 6191–6201 (2019). https://doi.org/10.1007/s00330-019-06163-2
22. Zhou, Z.H.: A brief introduction to weakly supervised learning. Natl. Sci. Rev. **5**(1), 44–53 (2017)

Boundary Distance Loss for Intra-/Extra-meatal Segmentation of Vestibular Schwannoma

Navodini Wijethilake[1(✉)], Aaron Kujawa[1], Reuben Dorent[1], Muhammad Asad[1], Anna Oviedova[2], Tom Vercauteren[2], and Jonathan Shapey[1,2]

[1] School of BMEIS, King's College London, London, UK
navodini.wijethilake@kcl.ac.uk
[2] Department of Neurosurgery, King's College Hospital, London, UK

Abstract. Vestibular Schwannoma (VS) typically grows from the inner ear to the brain. It can be separated into two regions, intrameatal and extrameatal respectively corresponding to being inside or outside the inner ear canal. The growth of the extrameatal regions is a key factor that determines the disease management followed by the clinicians. In this work, a VS segmentation approach with subdivision into intra-/extra-meatal parts is presented. We annotated a dataset consisting of 227 T2 MRI instances, acquired longitudinally on 137 patients, excluding post-operative instances. We propose a staged approach, with the first stage performing the whole tumour segmentation and the second stage performing the intra-/extra-meatal segmentation using the T2 MRI along with the mask obtained from the first stage. To improve on the accuracy of the predicted meatal boundary, we introduce a task-specific loss which we call Boundary Distance Loss. The performance is evaluated in contrast to the direct intrameatal extrameatal segmentation task performance, i.e. the Baseline. Our proposed method, with the two-stage approach and the Boundary Distance Loss, achieved a Dice score of 0.8279 ± 0.2050 and 0.7744 ± 0.1352 for extrameatal and intrameatal regions respectively, significantly improving over the Baseline, which gave Dice score of 0.7939 ± 0.2325 and 0.7475 ± 0.1346 for the extrameatal and intrameatal regions respectively.

1 Introduction

Vestibular schwannomas (VS) are benign intracranial tumours that arise from the insulating Schwann cells of the vestibulocochlear nerve. Typically they begin to grow within the internal auditory canal, often expanding the internal auditory meatus (IAM) and extending medially towards the brainstem, causing symptoms ranging from headache, hearing loss and dizziness to speech and swallowing difficulties as well as facial weakness. VS accounts for 8% of intra-cranial tumours, and is considered the most common nerve sheath tumour in adults [11].

A. Abdulkadir et al. (Eds.): MLCN 2022, LNCS 13596, pp. 73–82, 2022.
https://doi.org/10.1007/978-3-031-17899-3_8

The decisions of tumour management, which can be either active treatment procedures (surgery or radiotherapy) or wait-and-scan strategy, are taken based on the growth patterns of the tumour [12]. Irrespective of the timing or type of treatment, surveillance of the tumour following treatment is required, where consistent and reliable measurements of the tumour are necessary to estimate tumour size and behaviour [2]. According to the guidelines for reporting results in Acoustic Neuroma, the intrameatal and extrameatal portions of the tumour are required to be distinguished clearly and the largest extrameatal diameter should be used to report the size of the tumour [6]. Therefore, this specific segmentation of intrameatal and extrameatal regions is an important task in providing a reliable routine for reporting and analysing the growth of VS.

Routinely, the extraction of largest extrameatal dimension on the axial plane is performed manually by clinicians as there is no automated framework available in current clinical settings. Thus, the measurements extracted, are prone to subjective variability and also, it is a tedious, labour intensive task [13]. Therefore, it is essential to develop an AI framework for intra-/extra-meatal segmentation which we can later integrate into clinical settings along with automated size measurement extraction. With this the reproducibility and repeatability can be ensured. Nonetheless, according to previous studies, the volumetric measures are more repeatable than linear measurements extracted from small VS [8,15]. These volumetric measurements can be reliably extracted using the 3D tumour segmentations.

Related Work. Contrast-enhanced T1 weighted MRI (ceT1) and T2 weighted MRI are frequently utilized for VS management. Several AI approaches have been proposed for VS whole tumour segmentation within the past few years. Shapey et al. [14] have achieved a Dice score of 0.9343 and 0.8825 with ceT1 and T2 modalities respectively, using a 2.5D convolutional neural network (CNN). Dorent et al. [3] proposed the CrossMoDA computational challenge for VS and cochlea segmentation using T2 MRI with domain adaptation. In CrossMoDA the best performing method reached a Dice score of 0.8840 for the VS structure.

Using this approach, the authors emphasise how T2 weighted imaging may be routinely utilised for surveillance imaging, increasing patient safety by reducing the need to use gadolinium contrast agents. T2 weighted MRIs are also identified as 10 times more cost-effective than ceT1 imaging [1,2]. In a recent study, Neve et al. [10] have reported a Dice score of 0.8700 using T2 weighted MRI on an independent test set, where they have also used the whole tumour segmentation to distinguish the intrameatal and extrameatal regions of VS.

Multi-stage approaches have been proposed to hierarchically segment substructures of brain gliomas [16]. The authors claim that cascades can reduce overfitting by focusing on specific regions at each stage while reducing false positives. However, such approaches have not been used for VS-related tasks.

Boundary-based segmentation losses have been developed to address the issues associated with the overlap-based losses in highly unbalanced segmentation problems [7]. Hatamizadeh et al. [4] have proposed a deep learning

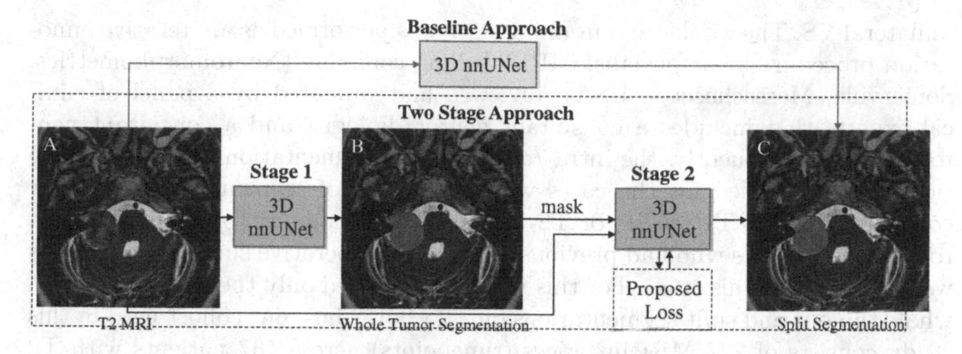

Fig. 1. The outline of the segmentation task. **A** shows an axial slice of the T2 MRI volume. **B** shows the output of stage 1, i.e. the whole tumour segmentation. **C** shows the output of stage 2, the split segmentation (red label: extrameatal region & green label: intrameatal region. The two-stage approach is shown within the dotted box and the baseline approach is highlighted in blue above the dotted box. The proposed loss function is used in stage 2. (Color figure online)

architecture, that consists of a separate module that learns the boundary information which aggregates an the edge aware loss, to the semantic loss. In [17], the Boundary Weighted Segmentation Loss (BWSL) combines distance map of the ground truth with the predicted probability map in order to make the network more sensitive to the boundary regions. Overall, existing literature on boundary losses seems focused on closed contours and a solution dedicated to specific boundary sections has not yet been adopted.

Contributions. In this work, we propose a two-stage approach, as illustrated in Fig. 1. The first stage performs the whole tumour segmentation and the second stage performs the intra-/extra-meatal segmentation using T2 weighted MRI along with the whole tumour mask obtained from stage 1. To the best of our knowledge, our study is the first to propose a fully automated learning based approach for intra-/extra-meatal segmentation of VS. We propose a new Boundary Distance Loss and demonstrate its advantage for learning the boundary between the intra-/extra-meatal regions accurately. We compare the performance of our staged approach and the novel loss function with a baseline, where the split segmentation is performed directly with the T2 weighted MRI volume without the staged approach. Additionally, we also compare the results of stage 2 of the two-stage approach with and without the proposed loss.

2 Methods

Dataset. The original dataset consists of MRI scans collected from 165 patients, whose initial MRI scanning was performed during the period February 2006 to January 2017. Further, the follow-up MRI scans were performed until September 2019. The patient cohort was older than 18 years and diagnosed with a single unilateral VS. The whole tumour annotation was performed as an iterative annotation procedure by a specialist MRI-labelling company (Neuromorphometrics, Somerville, Massachusetts, USA), reviewed and validated by a panel of clinical experts that includes a consultant neuroradiologist and a consultant neurosurgeon. Subsequently, the intra-/extra-meatal segmentation (split segmentation) was performed for the cases which consisted of segmentations of either contrast-enhanced T1 (ceT1) or T2 weighted MRI modality, by an expert neurosurgeon. Patients who had previously undergone operative surgical treatments were excluded at this stage. For this work, we included only the timepoints with whole tumour and split segmentations on T2 MRI. Thus, our cohort used in this study consists of 227 MRI instances (timepoints) across 137 patients with T2 MRI. The dataset is split into training, validation and testing sets, each with 195, 32 and 56 instances, respectively. We ensure that all timepoints belonging to a single patient are assigned to the same set, i.e. training or validation or testing set.

This study was approved by the NHS Health Research Authority and Research Ethics Committee (18/LO/0532). Because patients were selected retrospectively and the MR images were completely anonymised before analysis, no informed consent was required for the study.

Training: Baseline Approach. For our baseline, we make use of the default 3D full resolution UNet of the nnU-Net framework (3D nnU-Net) [5] to obtain the intra-/extra-meatal segmentation with the T2 MRI modality as the input for the network. A weighted cross entropy and Dice score losses are used for training.

Training: Two-Stage NnU-Net Approach. Similarly, we have used the 3D nnU-Net in two-stages sequentially in order to optimize the split segmentation task. All the training was performed on the NVIDIA Tesla V100 GPUs. Each model was trained for 1000 epochs, and the best performing model during validation was used to obtain the inference results.

Stage 1: Whole tumour segmentation. In stage 1, the 3D nnU-Net is used to segment the whole tumour region of VS with the T2 MRI as input. The combined loss of Cross Entropy and Dice score is used in this stage.

Stage 2: Intra-/Extra-Meatal Segmentation. In stage 2, the whole tumour mask, in addition to the T2 MRI, is given to the 3D nnU-Net to segment the tumour into intrameatal and extrameatal regions. For training, the manually annotated masks have been used but during inference, the predicted masks from stage 1 have been used for evaluation. We use a combination of cross entropy, Dice loss and our proposed Boundary Distance Loss detailed below.

Boundary Distance Loss Function. We define $I \in \mathbb{R}^{H \times W \times D}$ as the T2 MRI volumes with height, width, depth of H, W, D. Any corresponding (probabilistic) binary label map is denoted by $L_{\text{label}} \in \mathbb{R}^{H \times W \times D}$. The goal of this proposed loss function is to learn the deviation in the prediction from the actual boundary of intrameatal and extrameatal tumour regions.

Let's assume that for the three classes (background, intrameatal region and extrameatal region) the prediction map, i.e. the softmax probability maps of the neural network, is denoted by $P = [P_0, P_1, P_2]$. The spatial gradients $\nabla_x P_i$, $\nabla_y P_i$, $\nabla_z P_i$ in the x, y, and z directions provide spatial gradient magnitudes:

$$|\nabla P_i| = \sqrt{\nabla_x P_i^2 + \nabla_y P_i^2 + \nabla_z P_i^2}, \quad i \in \{0, 1, 2\} \tag{1}$$

The boundary between the intra-/extra-meatal regions should feature as edges in both corresponding label probability maps. As such, we multiply the magnitudes of the spatial gradients from the corresponding probability maps to achieve an intra-/extra-meatal boundary detector $B_P^{1,2}$ from our network predictions:

$$B_P^{1,2} = |\nabla P_1| \cdot |\nabla P_2| \tag{2}$$

Let $L_B \in \mathbb{R}^{H \times W \times D}$ denote the ground-truth one-hot encoded binary boundary map between the intra- and extra-meatal tumour regions. The Euclidean distance map ϕ_B from this ground-truth boundary is defined as,

$$\phi_B(u) = \begin{cases} 0, & L_B(u) = 1 \\ \inf\limits_{v | L_B(v) = 1} \|u - v\|_2, & L_B(u) = 0 \end{cases} \tag{3}$$

The PyTorch implementation of the Euclidean distance map is retrieved from the FastGeodis package[1]. To promote boundary detections that are close to the true boundary while being robust to changes far away from the true boundary, we compute the average D_{ϕ_B} negative scaled exponential distance between the detected boundary points and the ground truth:

$$D_{\phi_B} = \frac{\sum_u B_P^{1,2}(u) \exp\left(-\phi_B(u)/\tau\right)}{\sum_u B_P^{1,2}(u)} \tag{4}$$

where τ is a hyperparameter acting as a temperature term. We finally compute our Boundary Distance Loss $\mathcal{L}_B$ by taking the negative logarithm of D_{ϕ_B}

$$\mathcal{L}_B = -\log\left(D_{\phi_B}\right) \tag{5}$$

To define our complete loss $\mathcal{L}$ for training, we combine $\mathcal{L}_B$ with the cross-entropy ($\mathcal{L}_{CE}$) and the Dice loss ($\mathcal{L}_{DC}$) weighted by a factor γ:

$$\mathcal{L} = \mathcal{L}_{CE} + \mathcal{L}_{DC} + \gamma \mathcal{L}_B \tag{6}$$

---

[1] https://github.com/masadcv/FastGeodis.

Evaluation. Following the recent consensus recommendations on the selection of metrics for biomedical image analysis tasks [9], we evaluate our results using the Dice score as the primary overlap-based segmentation metric. Also following [9], the average symmetric surface distance (ASSD) metric is used to measure the deviation in the prediction, from the actual boundary of intrameatal and extrameatal tumour regions.

Table 1. Inference: Comparison of Dice score. BG: Background, EM: ExtraMeatal, IM: IntraMeatal, WT: Whole tumour, SD: Standard Deviation.

Method	γ		BG	EM	IM	WT
nnU-Net (Baseline)	–	Mean	0.9998	0.7939	0.7475	0.8813
	–	SD	0.0002	0.2325	0.1346	0.0888
Two-stage nnU-Net Approach						
Stage 1: WT Segmentation	–	Mean	0.9998	–	–	0.9039
	–	SD	0.0002	–	–	0.0470
Stage 2: Split Segmentation	0	Mean	0.9996	0.8068	0.7357	0.9026
		SD	0.0004	0.2231	0.1522	0.0478
	0.01	Mean	0.9995	0.8072	0.7368	0.9026
		SD	0.0004	0.2213	0.1524	0.0651
	0.05	Mean	0.9995	0.8087	0.7469	0.9027
		SD	0.0004	0.2243	0.1431	0.0651
	0.1	Mean	0.9996	0.8155	0.7655	0.9027
		SD	0.0004	0.2176	0.1329	0.0476
	0.5	Mean	0.9995	**0.8279**	**0.7744**	0.9025
		SD	0.0004	0.2050	0.1352	0.0478

Table 2. Inference: Comparison of ASSD metric; $\mathbf{p}_X$: X^{th} percentile of the ASSD metric distribution in percentage.

Method	γ	Median	$\mathbf{p}_{75}$	$\mathbf{p}_{25}$
nnU-Net (Baseline)	–	0.8384	1.1241	0.4326
Two-stage nnU-Net Approach				
Stage 1: Whole tumour Segmentation	–	–	–	
Stage 2: Split Segmentation	0	0.8064	1.3529	0.5690
	0.01	0.8020	1.2303	0.5351
	0.05	0.7024	1.1350	0.4830
	0.1	0.7602	0.9597	0.4546
	0.5	**0.5417**	0.9586	0.4181

3 Results

The quantitative comparison of the Dice score from the inference phase is given in Table 1. The Baseline approach gave a Dice score of 0.7939 ± 0.2325 and 0.7475 ± 0.1346 for extrameatal and intrameatal regions. This was improved significantly ($p < 0.01$) to 0.8068 ± 0.2231 and 0.7357 ± 0.1522 respectively for extrameatal and intrameatal regions, with the two-stage approach with combined loss of Cross Entropy and Dice Lose. This performance was further enhanced significantly ($p < 0.0001$) with the proposed Boundary Distance Loss ($\mathcal{L}_B$), which gave a dice score of 0.8279 ± 0.2050 and 0.7744 ± 0.1352 respectively for extrameatal and intrameatal regions with $\gamma = 0.5$.

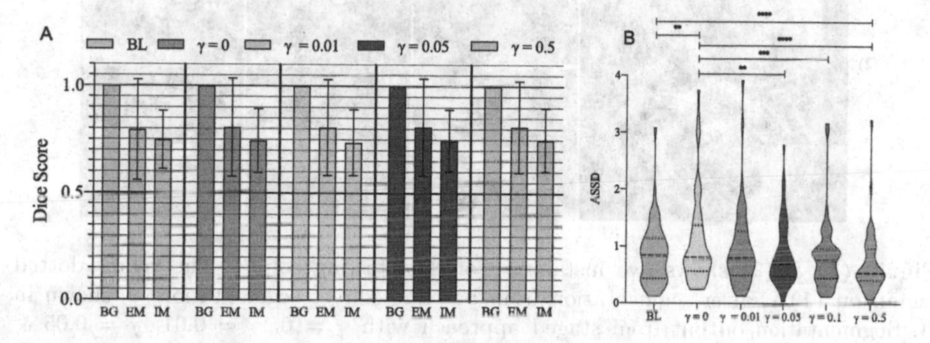

Fig. 2. **A** shows the Dice score distribution for the Background (BG), Extrameatal (EM) and Intrameatal (IM) for Baseline method, two-stage approach with $\gamma = 0$, $\gamma = 0.01$, $\gamma = 0.05$ and $\gamma = 0.5$ respectively. **B** illustrates the ditribution of the ASSD metric for the Baseline method, two-stage approach with $\gamma = 0$, $\gamma = 0.01$, $\gamma = 0.05$ and $\gamma = 0.5$ respectively.

Figure 2 shows the distribution of Dice score and the ASSD metric for the Baseline method and two-stage approach for $\gamma = 0$, $\gamma = 0.01$, $\gamma = 0.05$ and $\gamma = 0.5$ respectively. We further performed a two-sided Wilcoxon matched pairs signed-rank test, in which each distribution is considered significantly different from the other distribution when $p < 0.05$.

In Fig. 3, we illustrate qualitative results, i.e. two instances from the testing cohort. The instance (1) gave a high ASSD with baseline and with $\gamma = 0$ for the two-stage approach, compared to the other results obtained with the proposed Boundary Distance Loss. The instance (2) had given a high ASSD metric with the baseline and two-stage approach with the $\gamma = 0$ had given the ASSD value of inf, as it had not distinguished the extrameatal region. The split segmentation obtained with the proposed loss had given comparatively lower ASSD values with $\gamma = 0.01$ and $\gamma = 0.5$ for this instance.

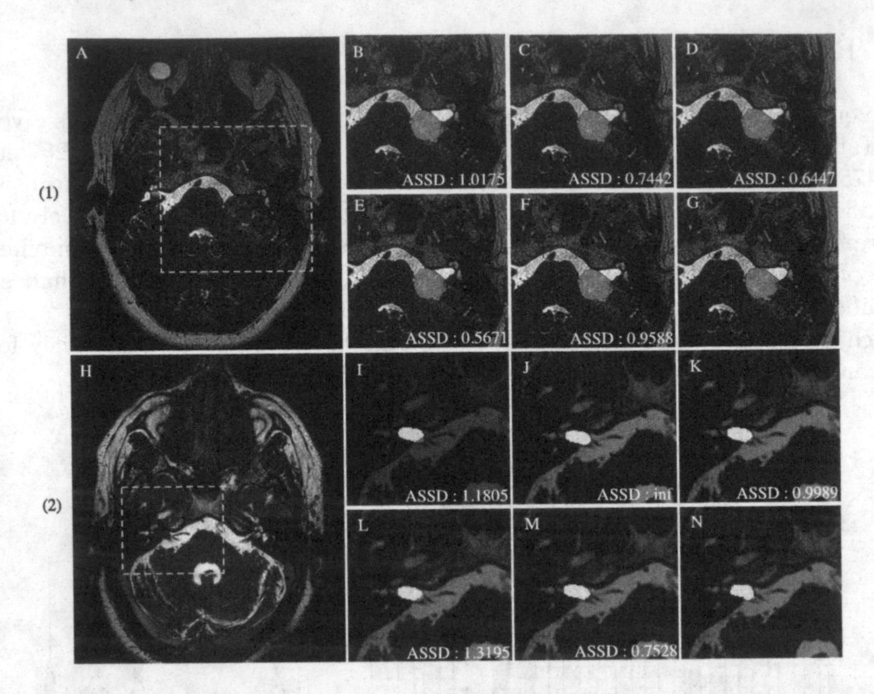

Fig. 3. **(1)** & **(2)** shows two instances from the testing cohort. The yellow dotted region on **(1)A**, after segmentation from baseline (direct segmentation) is shown in **B**. Segmentation output from staged approach with $\gamma = 0$, $\gamma = 0.01$, $\gamma = 0.05$ & $\gamma = 0.5$ are shown in **C**, **D**, **E**, **F** respectively. The ground truth is shown in **G**. Similarly, the yellow dotted region on **(2)H**, after segmentation from baseline is shown in **I**. Segmentation output from staged approach with $\gamma = 0$, $\gamma = 0.01$, $\gamma = 0.05$ & $\gamma = 0.5$ are shown in **J**, **K**, **L**, **M** respectively. The ground truth is shown in **N**. The corresponding ASSD metric is shown on each sub figure.

4 Discussion and Conclusion

In this work, we use a two-stage approach with a novel Boundary Distance Loss for intrameatal and extrameatal segmentation of VS. The two-stage approach with boundary loss shows a promising improvement over the baseline approach of direct intrameatal and extrameatal segmentation. The two-stage approach helps the model to focus on whole tumour segmentation at the first stage and then, with the mask from stage 1, model can learn the intrameatal and extrameatal separation boundary rigorously during the stage 2. Furthermore, the proposed loss function enhance the split boundary identification by learning the distance between the predicted and the target boundary.

The results indicate that a low γ weight, such as 0.01 or 0.05, on the Boundary Distance Loss does not significantly improve the performance over the zero weight on the proposed loss. However, a significant improvement could be seen with higher γ values such as 0.1 or 0.5. In the future, we will fine-tune the γ

hyper parameter more precisely to assign the most appropriate weight on the proposed Boundary Distance Loss.

For the training, with the proposed loss the training time for a single epoch increases approximately by two times. In addition, we observed that during inference, if we use the manually annotated mask instead of the mask obtained from the stage 1, the split segmentation improves further. Thus, we can assume if the stage 1 performance can be improved, the stage 2 performance can be enhanced. Another appropriate approach would be training the stage 2 with the predicted masks from the stage 1.

In conclusion, our method reveals the importance of learning the distance to the boundary in tasks that require distinguishing the boundary precisely. This improvement over the boundary is quite crucial, as it enhances the extraction of features, such as the largest extrameatal diameter from the extrameatal region. Our proposed loss can be used in similar applications that also require a boundary determination from a whole region segmentation.

Acknowledgement. The authors would like to thank Dr Andrew Worth and Gregory Millington for their contributions to the generation of the segmentation ground truth.

Disclosures. N. Wijethilake was supported by the UK Medical Research Council [MR/N013700/1] and the King's College London MRC Doctoral Training Partnership in Biomedical Sciences. This work was supported by Wellcome Trust (203145Z/16/Z, 203148/Z/16/Z, WT106882), EPSRC (NS/A000050/1, NS/A000049/1) and MRC (MC/PC/180520) funding. Additional funding was provided by Medtronic. TV is also supported by a Medtronic/Royal Academy of Engineering Research Chair (RCSRF1819/7/34). SO is co-founder and shareholder of BrainMiner Ltd., UK.

References

1. Coelho, D.H., Tang, Y., Suddarth, B., Mamdani, M.: MRI surveillance of vestibular schwannomas without contrast enhancement: clinical and economic evaluation. Laryngoscope **128**(1), 202–209 (2018)
2. Connor, S.E.: Imaging of the vestibular schwannoma: diagnosis, monitoring, and treatment planning. Neuroimag. Clin. **31**(4), 451–471 (2021)
3. Dorent, R., et al.: Crossmoda 2021 challenge: benchmark of cross-modality domain adaptation techniques for vestibular schwnannoma and cochlea segmentation. arXiv preprint arXiv:2201.02831 (2022)
4. Hatamizadeh, A., Terzopoulos, D., Myronenko, A.: End-to-end boundary aware networks for medical image segmentation. In: Suk, H.-I., Liu, M., Yan, P., Lian, C. (eds.) MLMI 2019. LNCS, vol. 11861, pp. 187–194. Springer, Cham (2019). https://doi.org/10.1007/978-3-030-32692-0_22
5. Isensee, F., Jaeger, P.F., Kohl, S.A., Petersen, J., Maier-Hein, K.H.: nnU-Net: a self-configuring method for deep learning-based biomedical image segmentation. Nat. Methods **18**(2), 203–211 (2021)
6. Kanzaki, J., Tos, M., Sanna, M., Moffat, D.A.: New and modified reporting systems from the consensus meeting on systems for reporting results in vestibular schwannoma. Otol. Neurotol. **24**(4), 642–649 (2003)

7. Kervadec, H., Bouchtiba, J., Desrosiers, C., Granger, E., Dolz, J., Ayed, I.B.: Boundary loss for highly unbalanced segmentation. In: International Conference on Medical Imaging with Deep Learning, pp. 285–296. PMLR (2019)
8. MacKeith, S., et al.: A comparison of semi-automated volumetric vs linear measurement of small vestibular schwannomas. Eur. Archiv. Oto-Rhino-Laryngol. **275**(4), 867–874 (2018)
9. Maier-Hein, L., et al.: Metrics reloaded: pitfalls and recommendations for image analysis validation. arXiv preprint arXiv:2206.01653 (2022)
10. Neve, O.M., et al.: Fully automated 3D vestibular schwannoma segmentation with and without gadolinium contrast: a multicenter, multivendor study. Radiol. Artif. Intell. e210300 (2022)
11. Ostrom, Q.T., Cioffi, G., Waite, K., Kruchko, C., Barnholtz-Sloan, J.S.: Cbtrus statistical report: primary brain and other central nervous system tumors diagnosed in the united states in 2014–2018. Neuro-oncology 23(Supplement_3), iii1–iii105 (2021)
12. Prasad, S.C., et al.: Decision making in the wait-and-scan approach for vestibular schwannomas: is there a price to pay in terms of hearing, facial nerve, and overall outcomes? Neurosurgery **83**(5), 858–870 (2018)
13. Shapey, J., et al.: Artificial intelligence opportunities for vestibular schwannoma management using image segmentation and clinical decision tools. World Neurosurg. **149**, 269–270 (2021)
14. Shapey, J., et al.: An artificial intelligence framework for automatic segmentation and volumetry of vestibular schwannomas from contrast-enhanced t1-weighted and high-resolution t2-weighted MRI. J. Neurosurg. **134**(1), 171–179 (2019)
15. Varughese, J.K., Breivik, C.N., Wentzel-Larsen, T., Lund-Johansen, M.: Growth of untreated vestibular schwannoma: a prospective study. J. Neurosurg. **116**(4), 706–712 (2012)
16. Wang, G., Li, W., Ourselin, S., Vercauteren, T.: Automatic brain tumor segmentation using cascaded anisotropic convolutional neural networks. In: Crimi, A., Bakas, S., Kuijf, H., Menze, B., Reyes, M. (eds.) BrainLes 2017. LNCS, vol. 10670, pp. 178–190. Springer, Cham (2018). https://doi.org/10.1007/978-3-319-75238-9_16
17. Zhu, Q., Du, B., Yan, P.: Boundary-weighted domain adaptive neural network for prostate MR image segmentation. IEEE Trans. Med. Imaging **39**(3), 753–763 (2019)

Neuroimaging Harmonization Using cGANs: Image Similarity Metrics Poorly Predict Cross-Protocol Volumetric Consistency

Veronica Ravano[1,2,3](✉), Jean-François Démonet[2], Daniel Damian[2], Reto Meuli[2], Gian Franco Piredda[1,2,3], Till Huelnhagen[2,3], Bénédicte Maréchal[1,2,3], Jean-Philippe Thiran[2,3], Tobias Kober[1,2,3], and Jonas Richiardi[2]

[1] Advanced Clinical Imaging Technology, Siemens Healthcare AG, Lausanne, Switzerland
veronica.ravano@epfl.ch

[2] Department of Radiology, Lausanne University Hospital and University of Lausanne, Lausanne, Switzerland

[3] LTS5, École Polytechnique Fédérale de Lausanne (EPFL), Lausanne, Switzerland

Abstract. Computer-aided clinical decision support tools for radiology often suffer from poor generalizability in multi-centric frameworks due to data heterogencity. In particular, magnetic resonance images depend on a large number of acquisition protocol parameters as well as hardware and software characteristics that might differ between or even within institutions. In this work, we use a supervised image-to-image harmonization framework based on a conditional generative adversarial network to reduce inter-site differences in T1-weighted images using different dementia protocols. We investigate the use of different hybrid losses including standard voxel-wise distances and a more recent perceptual similarity metric, and how they relate to image similarity metrics and volumetric consistency in brain segmentation. In a test cohort of 30 multiprotocol patients affected by dementia, we show that despite improvements in terms of image similarity, the synthetic images generated do not necessarily result in reduced inter-site volumetric differences, therefore highlighting the mismatch between harmonization performance and the impact on the robustness of post-processing applications. Hence, our results suggest that traditional image similarity metrics such as PSNR or SSIM may poorly reflect the performance of different harmonization techniques in terms of improving cross-domain consistency.

Keywords: Domain adaptation · MRI · Harmonization

1 Introduction

In neurological applications, computer-aided clinical decision support tools typically provide a volumetric estimation of anatomical brain regions (e.g., thalamus, hippocampus [1]) or detect focal abnormalities (e.g., white matter lesion [2], metastases [3]), often

---

Supplementary Information The online version contains supplementary material available at https://doi.org/10.1007/978-3-031-17899-3_9.

A. Abdulkadir et al. (Eds.): MLCN 2022, LNCS 13596, pp. 83–92, 2022.
https://doi.org/10.1007/978-3-031-17899-3_9

based on Magnetic Resonance Imaging (MRI). However, the scalability of such algorithms has been hindered so far by the heterogeneity that characterizes MRI data. In fact, when considering a given modality (e.g., T1-weighted, T2-weighted), the contrast, resolution and signal-to-noise ratio of an MR image typically depend on several acquisition parameters as well as software and hardware characteristics that can vary considerably between or even within institutions. Therefore, clinical decision support tools often fail to generalize properly to MRI data acquired in new "unseen" settings [4, 5].

To overcome this issue, domain adaptation techniques can be used to harmonize MR images from different sites, protocols and/or acquisition hardware to a common reference domain [6–9]. Supervised image-to-image translation techniques enable transformation of MRI scans from a source domain to anatomically equivalent images mimicking the acquisition in a reference domain by leveraging ground truth target images provided by paired acquisitions of the same subject. In this context, conditional Generative Adversarial Networks (cGANs) [9] have been widely used for cross-contrast [11, 12] (e.g., T1-weighted to T2-weighted) and cross-imaging modality translations [13–15] (e.g., MR to CT). Other common approaches for image-to-image translation include nonlinear regression models [16] that learn the intensity distribution of a target image, and the use of autoencoders, where a model is trained to disentangle the anatomical content of the image and its site-dependent style component [7, 17]. Compared to other harmonization techniques, supervised image-to-image models such as cGANs have the advantage of being insensitive to domain-specific biases that might occur in unpaired training datasets (e.g., systematic demographical differences between the two domains) and preserve anatomical consistency in the synthetic images by means of a voxel-wise reconstruction loss.

Most literature evaluates harmonization results using voxel-wise error metrics such as mean square error (MSE) and mean average error (MAE) [6, 15] or more high-level metrics including structural similarity metric (SSIM) and peak signal to noise ratio (PSNR) [6, 7, 12, 15, 16]. However, the relationship between these image similarity metrics and the ultimate cross-protocol consistency of volumetric estimates using a segmentation algorithm remains unclear, as does the impact of specific loss functions on specific post-processing applications.

In this work, we implement image-to-image translation of T1-weighted MRI scans using an adapted version of the pix2pix cGAN [10]. We investigate the use of both standard voxel-wise L1 and L2 distances, as well as a more recent perceptual similarity loss (LPIPS) [18]. We also evaluate histogram matching to the reference domain. We estimate the harmonization performance in terms of several image similarity metrics and consistency of volumetric estimates of brain regions in thirty patients affected with dementia, which were acquired with two different protocols. Brain volumes were estimated using MorphoBox [1], a brain segmentation algorithm prototype optimized for T1-weighted images acquired with ADNI-compatible [19] protocols.

2 Methods

2.1 Dataset

Seventy-four patients (70.2 ± 11.8 years old, 36 females) scanned for workup of cognitive decline were recruited in a study approved by the local ethics committee and provided written consent to participate in the study. T1-weighted magnetization-prepared rapid gradient echo (MPRAGE) images were acquired with two distinct parameter sets during the same session without patient repositioning. The acquisition protocol parameters used are presented in Table 1.

Table 1. Acquisition protocols used to acquire 3D MPRAGE images

	Protocol 1	Protocol 2
Scanner	MAGNETOM Prisma 3T (Siemens Healthcare, Erlangen, Germany)	
TR/TE/TI [ms]	2300/2.98/900	1930/2.36/972
Resolution [mm^3]	$1 \times 1 \times 1.1$	$0.87 \times 0.87 \times 0.9$
Flip Angle [°]	9	8
Pixel readout bandwidth [Hz/ms]	240	200

MR images were corrected for intensity non-uniformities using the N4 algorithm [20]. Subsequently, the images from Protocol 2 were spatially registered to the equivalent image acquired with Protocol 1 for each patient using Elastix [21] with an affine transformation followed by linear interpolation.

Protocol 1 corresponds to the ADNI standard [19], which is the recommended acquisition protocol for the MorphoBox segmentation algorithm; it was therefore defined to be the reference domain for our image-to-image translation, and brain segmentations derived from Protocol 1 were defined as ground truth.

2.2 Harmonization Framework

We based our cGAN implementation on the pix2pix model proposed in [10], but using a custom ResNet generator where transpose convolution was replaced with an upsampling operation followed by a convolution to reduce checkerboard artefacts (code available here: https://gitlab.com/acit-lausanne/harmonization_cgan.git). For each MRI volume used for training, 60 central slices were extracted in each orientation (axial, coronal and sagittal) and stacks of three consecutive slices were fed to the network as separate channels. For testing, synthetic images were first generated for each orientation, and the voxel-wise average across orientations was then computed.

To investigate the impact of hybrid losses on the accuracy of a harmonization cGAN model, we compared three distinct models with different loss components. We investigated the standard L1 norm capturing voxel-wise differences, the L2 norm for its increasing sensitivity to larger errors and the LPIPS similarity for its ability to capture more global perceptual similarity. Previous work showed the advantages of combining error measures together into a hybrid loss [22, 23]. Each loss component was weighted by a factor λ that was determined experimentally by considering the amplitude of each loss component during training. The resulting hybrid loss $\mathcal{L}_G$ can be formulated as follows, with $\hat{y} = f(y, w)$ being the synthetic image resulting from the network f given its weights w, n being the number of voxels in the image and $\mathcal{L}_{adv}$ the adversarial loss.

$$\mathcal{L}_G = \lambda_{L1} \sum_i^n \frac{|\widehat{y}_i - y_i|}{n} + \lambda_{L2} \sum_i^n \frac{(\widehat{y}_i - y_i)^2}{n} + \lambda_{LPIPS} LPIPS(\hat{y}, y) + \mathcal{L}_{adv} \quad (1)$$

where $\lambda_{L1} = 1000, \lambda_{L2} = 10000$ *and* $\lambda_{LPIPS} = 500$.

Three folds were defined, each of which included 64 training patients and ten test patients. Bayesian hyperparameter optimization was performed in each fold with Optuna [24], using 60 patients for training and four for validation. The network defined by the best hyperparameter set was selected in each fold, retrained on 64 training patients and its accuracy estimated on the testing patients. The network had 57.2 M trainable parameters and was trained on a NVIDIA Tesla V100 32 GB GPU (hyperparameters provided in Supplementary Table 1).

MorphoBox tissue classification algorithm requires accurate guesses of the cerebrospinal fluid (CSF), gray and white matter tissue mean intensities. The latter are assessed effectively by detecting the three zero-crossings of the smoothed histogram first derivative of the skull-stripped T1-weighted image. Therefore, systematic site-related differences in the intensity histograms are likely to introduce a bias in brain segmentation. Thus, we also tested histogram matching (ITK [25] implementation, with excluding background voxels, 100 histogram levels and 15 matching quantile values) to a reference ADNI image, selected from the ADNI standardized analysis set described in [19] (female aged 71 years old, cognitive normal), as an alternative to the cGAN.

2.3 Statistical Analysis

The accuracy of the harmonized images was evaluated in terms of similarity compared to the reference Protocol 1 images and in terms of the cross-protocol volumetric estimation consistency using MorphoBox.

To this end, MAE and MSE errors alongside more global image similarity metrics SSIM [26] and PSNR were computed between the harmonized image and the respective reference. Further, we compared the intensity histograms and evaluated their proximity by computing the Wasserstein distance (WD) [27]. A brain mask was used to consider only the voxels in the brain.

The post-harmonization similarity metrics were compared to the original metrics with a paired one-sided Wilcoxon test (hypothesizing harmonization will increase similarity), corrected for multiple comparisons (Benjamini-Hochberg False Discovery Rate, nominal significance level 0.05).

Bland-Altman analyses using relative differences were performed to compare the agreement between the different harmonization techniques and the baseline in terms of brain volume estimations using MorphoBox. For each brain region and each method, we evaluated the range of agreement and the absolute bias.

3 Results

3.1 Image Similarity

When computed on the original images, the L1 and L2 losses were highly correlated (Spearman $\rho = 98.8\%$), suggesting that the combination of the two would not improve the harmonization accuracy. In contrast to this, LPIPS was only weakly correlated with the voxel-wise L1 distance ($\rho = 22.0\%$), suggesting that complementary information could possibly be brought by such a perceptual metric.

Table 2. Similarity and error measures computed between the reference Protocol 1 image and the original or harmonized Protocol 2 equivalent scan for each testing patient, presented as mean ± standard deviation over the 30 test patients. Arrows indicate the desired trend for higher similarity (↓: lower value for higher similarity, ↑: higher value for higher similarity). MAE: mean average error, MSE: mean squared error, SSIM: structural similarity metric, PSNR: peak signal to noise ratio, WD: Wasserstein distance. Best values are highlighted in **bold**.

Model	MAE ↓	MSE ↓	WD ↓	SSIM ↑	PSNR ↑
baseline	128 ± 18.0	18.56 ± 4.80x10^3	127 ± 18.1	0.75 ± 0.04	14.5 ± 1.5
hist. Matching [25]	52.9 ± 29.4	4.46 ± 4.48x10^3	51.8 ± 30.5	**0.97 ± 0.02**	**24.9 ± 4.9**
cGAN (L1)	55.8 ± 10.3	4.19 ± 1.37x10^3	53.1 ± 11.9	0.84 ± 0.03	21.0 ± 2.3
cGAN (L1 + L2)	53.4 ± 12.4	4.00 ± 1.63x10^3	49.6 ± 14.7	0.84 ± 0.04	21.3 ± 2.7
cGAN (L1 + LPIPS)	**52.4 ± 13.0**	**3.84 ± 1.70x10^3**	**49.1 ± 14.9**	0.85 ± 0.04	21.6 ± 2.7

The accuracy of our harmonization models was first evaluated in terms of image similarity. The similarity measures computed between the harmonized images and the reference are reported in Table 2. Overall, harmonized images showed significantly higher image similarity compared to original protocols in all cases ($p_{FDR} < 0.05$). The lowest voxel-wise errors (MAE and MSE) and WD were observed for harmonized images with the cGAN trained with a combination of L1 and LPIPS loss, whilst the highest SSIM and PSNR were achieved for histogram matching. No significant improvement was observed for cGAN models compared to histogram matching. Globally, cGAN harmonization decreased the standard deviation of the similarity measures compared to baseline and histogram matching, therefore reflecting an improved stability across testing patients.

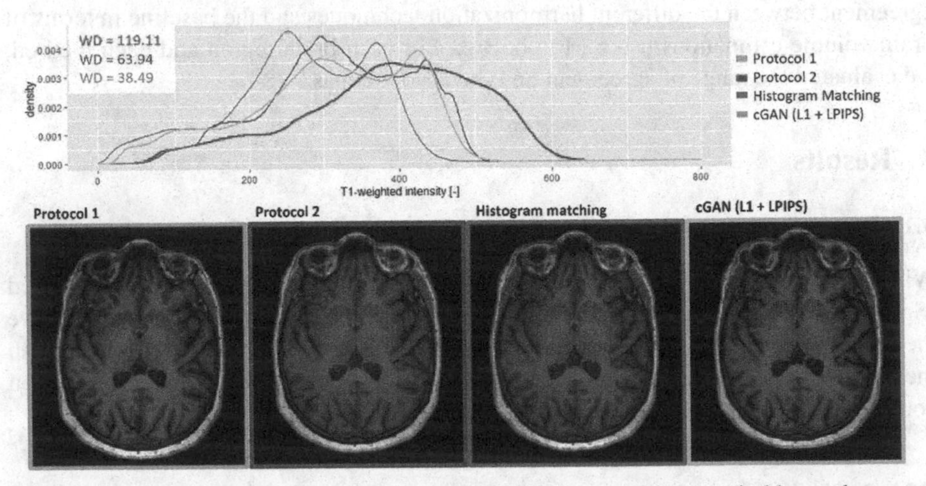

Fig. 1. Intensity histograms of original Protocol 1 and Protocol 2 images (in blue and orange, respectively), histogram matched images (in red) and harmonized images using a cGAN with L1 and LPIPS loss (grey), for one example patient. The Wasserstein distance (WD) between the reference Protocol 1 image and each other scan is reported and color-coded accordingly. The resulting contrasts are also shown below for the same patient (Color figure online).

Intensity histograms for one example patient are shown in Fig. 1 alongside the original and harmonized contrasts, where each main brain structure (cerebrospinal fluid, gray matter and white matter tissue) forms an intensity peak. Whilst three peaks could easily be observed in the image acquired with Protocol 1, the lower contrast between white and gray matter is reflected by a flatter histogram in the higher intensity ranges for Protocol 2.

After harmonization, the third intensity peak is recovered, thereby revealing an improved contrast, that can also be appreciated in the subcortical structures of the obtained MR images shown in Fig. 2. Images harmonized with the cGAN result in more homogeneous intensities within the same brain tissue compared to histogram matching, assumingly helping the segmentation algorithm.

3.2 Volumetric Consistency

The performance of the harmonization was also evaluated in terms of MorphoBox volumetry consistency before and after protocol harmonization. As brain tissue classification initialization approach is known to be sensitive to the modes of the intensity histogram of an image, we expected the images with lower WD to be associated with a more reliable volumetry estimation.

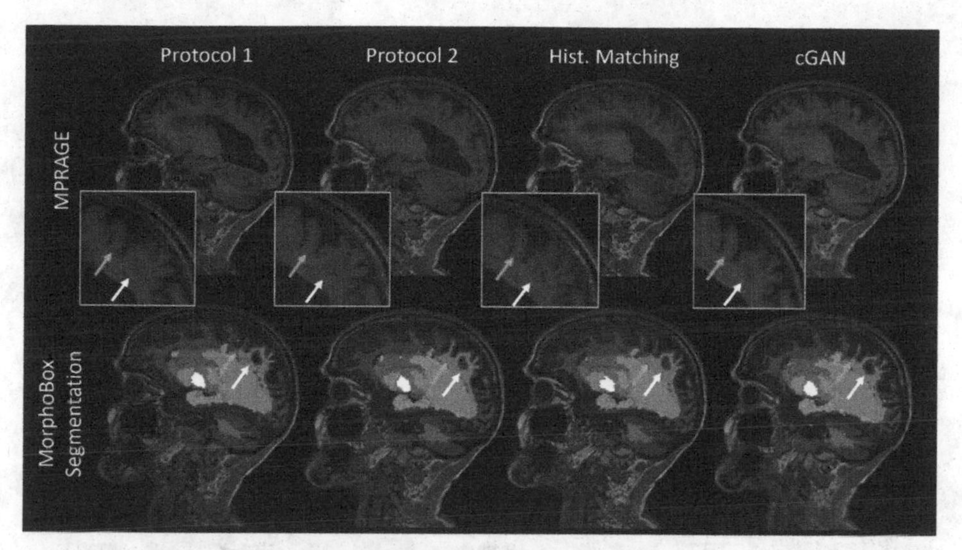

Fig. 2. Resulting MPRAGE contrasts and MorphoBox segmentation for an example patient, both prior and after harmonization using histogram matching and a deep learning cGAN model employing a combined L1 and LPIPS loss. The red arrows point at differences in the segmentation of the frontal and parietal lobe, whereas the yellow arrows highlight the segmentation of the gray matter and CSF in a particular gyrus (Color figure online).

Figure 2 shows the original and harmonized contrasts for an example patient, together with representative slices showing the resulting MorphoBox segmentations. The red arrows highlight differences in the boundaries between the frontal and parietal lobes, whereas the yellow arrows show differences in the gray matter segmentation in a particular gyrus. Overall, the segmentation results obtained from harmonized images were closer to the reference, particularly for the deep learning model. Notably, the image volumes resulting from histogram matching showed inhomogeneous intensities, especially in the white matter. In contrast, the synthetic images obtained with the deep learning harmonization model had an increased white/gray matter contrast but were more blurred.

Figure 3 reports the limits of agreement (left) computed from Bland-Altman analysis, the extend of the range of agreement (top right) and the absolute bias (bottom right) for each brain region using each harmonization technique compared to the Protocol 1 images. Overall, brain volumes estimated from harmonized images using L1 and LPIPS loss (in green) were in stronger agreement with original volumes compared to the baseline (in red). Despite a limited effect size, the range of agreement was lower for most of the regions, whereas no substantial difference could be observed in the bias. The other harmonization models (in grey) resulted in comparable agreement with the baseline. Despite being associated with the highest SSIM and PSNR values, histogram matching did not allow to substantially improve the robustness of volumetric estimates.

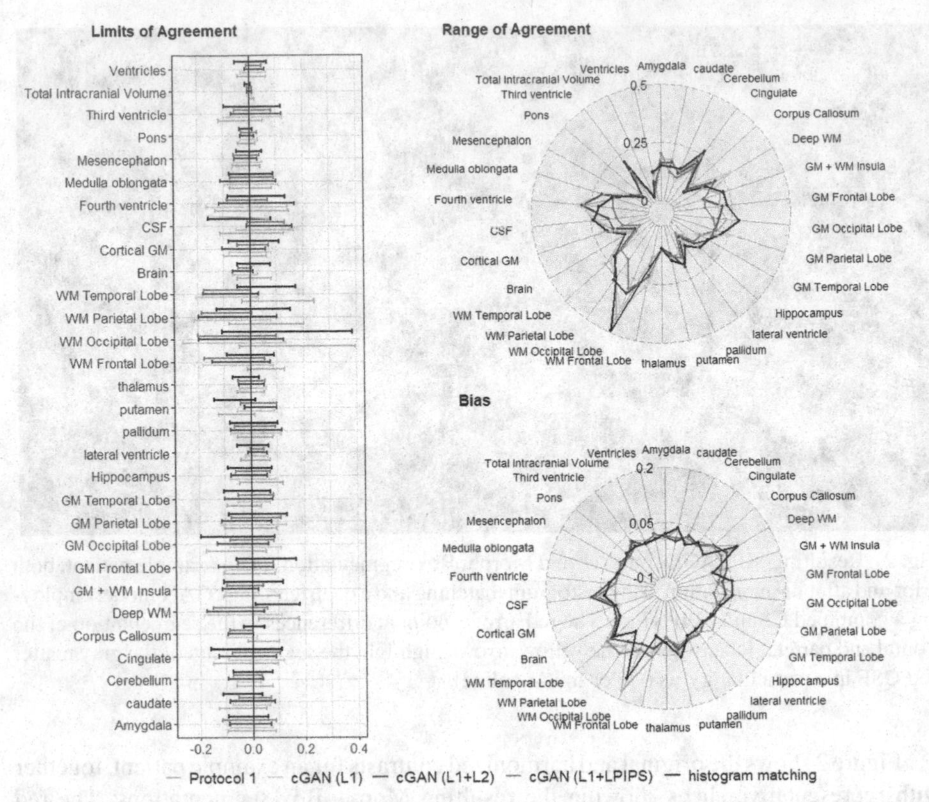

Fig. 3. Summary of Bland-Atlman analysis for brain volume estimates using MorphoBox from different harmonized images compared to Protocol 1. (Left) Limits of agreement computed for each brain region with a 95% confidence interval. (Top Right) Range of agreement for each brain region. (Bottom Right) Absolute bias for each brain region (Color figure online).

4 Discussion

We investigated the use of different tailored losses for a deep learning harmonization model based on a cGAN. We observed that the addition of a L2 component to a standard L1 loss used in the original pix2pix implementation did not improve the accuracy of the harmonization. The high correlation that was observed between the two measures might explain this observation. In contrast to that, the use of a perceptual metric, LPIPS, improved the harmonization results compared to standard L1 loss in terms of morphometry. Importantly, compared to matching intensity histograms to a reference ADNI subject, the cGAN is robust to differences in brain volumes and head size.

Overall, deep learning harmonization models showed slightly improved voxel-wise similarity metrics and WD between intensity histograms compared to histogram matching. In contrast, histogram-matched images resulted in higher similarity in terms of SSIM and PSNR. However, the differences were not statistically significant.

Our findings suggest that voxel-wise data consistency losses typically used in image-to-image translation models should be complemented with other metrics that better capture more global perceived similarity in MR images, such as LPIPS to improve the accuracy of the synthetically generated data.

Our findings also highlight a mismatch between image similarity metrics and volumetric consistency, suggesting that judging harmonization impact on the consistency of a given application (here, volumetry) based on image similarity bears risks: several image similarity metrics may favor a harmonization method that ultimately results in lower downstream performance, or indeed no difference in performance. As a consequence, one should consider that studies evaluating the performance of harmonization techniques solely based on image similarity metrics [15, 16] do not necessarily reflect an improvement in consistency of downstream post-processing analyses. Further, the design of the optimal reconstruction loss for harmonization purposes should take into consideration the intended clinical application and its sensitivity to different similarity measures, therefore challenging the concept of one harmonization fitting all applications.

Acknowledgments. This work was co-financed by Innosuisse (Grant 43087.1 IP-LS).

References

1. Schmitter, D., et al.: An evaluation of volume-based morphometry for prediction of mild cognitive impairment and Alzheimer's disease. NeuroImage Clin. **7**, 7–17 (2015). https://doi.org/10.1016/j.nicl.2014.11.001
2. Fartaria, M.J., et al.: Automated detection of white matter and cortical lesions in early stages of multiple sclerosis. J. Magn. Reson. Imaging **43**, 1445–1454 (2016). https://doi.org/10.1002/jmri.25095
3. Grøvik, E., Yi, D., Iv, M., Tong, E., Rubin, D., Zaharchuk, G.: Deep learning enables automatic detection and segmentation of brain metastases on multisequence MRI. J. Magn. Reson. Imaging **51**, 175–182 (2020). https://doi.org/10.1002/jmri.26766
4. Haller, S., et al.: Basic MR sequence parameters systematically bias automated brain volume estimation. Neuroradiology **58**(11), 1153–1160 (2016). https://doi.org/10.1007/s00234-016-1737-3
5. Hays, S.P., Zuo, L., Carass, A., Prince, J. Evaluating the impact of MR image contrast on whole brain segmentation. In: Medical Imaging 2022: Image Processing, vol. 12032, pp. 122–126. SPIE, April 2022
6. Dewey, B.E., et al.: DeepHarmony: a deep learning approach to contrast harmonization across scanner changes. Magn. Reson. Imaging **64**, 160–170 (2019). https://doi.org/10.1016/j.mri.2019.05.041
7. Dewey, B.E., et al.: A disentangled latent space for cross-site MRI harmonization. In: Martel, A.L., et al. (eds.) MICCAI 2020. LNCS, vol. 12267, pp. 720–729. Springer, Cham (2020). https://doi.org/10.1007/978-3-030-59728-3_70
8. Guan, H., Liu, Y., Yang, E., Yap, P.T., Shen, D., Liu, M.: Multi-site MRI harmonization via attention-guided deep domain adaptation for brain disorder identification. Med. Image Anal. **71**, 102076 (2021). https://doi.org/10.1016/j.media.2021.102076
9. Mirzaalian, H., et al.: Harmonizing diffusion MRI data across multiple sites and scanners. In: Navab, N., Hornegger, J., Wells, W.M., Frangi, A.F. (eds.) MICCAI 2015. LNCS, vol. 9349, pp. 12–19. Springer, Cham (2015). https://doi.org/10.1007/978-3-319-24553-9_2
10. Knyaz, V.A., Kniaz, V.V., Remondino, F.: Image-to-voxel model translation with conditional adversarial networks. In: Leal-Taixé, L., Roth, S. (eds.) ECCV 2018. LNCS, vol. 11129, pp. 601–618. Springer, Cham (2019). https://doi.org/10.1007/978-3-030-11009-3_37

11. Yang, Q., Li, N., Zhao, Z., Fan, X., Chang, E.I.C., Xu, Y.: MRI cross-modality image-to-image translation. Sci. Rep. **10**, 1–18 (2020). https://doi.org/10.1038/s41598-020-60520-6
12. Dar, S.U.H., Yurt, M., Karacan, L., Erdem, A., Erdem, E., Cukur, T.: Image synthesis in multi-contrast MRI with conditional generative adversarial networks. IEEE Trans. Med. Imaging **38**, 2375–2388 (2019). https://doi.org/10.1109/TMI.2019.2901750
13. Kaiser, B., Albarqouni, S.: MRI to CT Translation with GANs. (2019)
14. Jung, M.M., Berg, B. Van Den, Postma, E., Huijbers, W.: Inferring PET from MRI with pix2pix. In: Benelux Conference on Artificial Intelligence, pp. 1–9 (2018)
15. Armanious, K., et al.: MedGAN: medical image translation using GANs. Comput. Med. Imaging Graph. **79** (2020). https://doi.org/10.1016/j.compmedimag.2019.101684
16. Jog, A., Carass, A., Roy, S., Pham, D.L., Prince, J.L.: Random forest regression for magnetic resonance image synthesis. Med. Image Anal. **35**, 475–488 (2017). https://doi.org/10.1016/j.media.2016.08.009
17. Moyer, D., Ver Steeg, G., Tax, C.M.W., Thompson, P.M.: Scanner invariant representations for diffusion MRI harmonization. Magn. Reson. Med. **84**, 2174–2189 (2020). https://doi.org/10.1002/mrm.28243
18. Zhang, R., Isola, P., Efros, A.A., Shechtman, E., Wang, O.: The unreasonable effectiveness of deep features as a perceptual metric. Cvpr2018 **13** (2018)
19. Wyman, B.T., et al.: Standardization of analysis sets for reporting results from ADNI MRI data. Alzheimer's Dement. **9**, 332–337 (2013). https://doi.org/10.1016/j.jalz.2012.06.004
20. Tustison, N.J., Avants, B.B., Cook, P.A., Gee, J.C.: N4ITK : improved N3 bias correction with robust b-spline approximation. In: Proceedings of ISBI 2010, pp. 708–711 (2010)
21. Klein, S., Staring, M., Murphy, K., Viergever, M.A., Pluim, J.: elastix: a toolbox for intensity-based medical image registration. IEEE Trans. Med. Imaging. **29**, 196–205 (2010). https://doi.org/10.1109/TMI.2009.2035616
22. Grosenick, L., Klingenberg, B., Katovich, K., Knutson, B., Taylor, J.E.: Interpretable whole-brain prediction analysis with GraphNet. Neuroimage **72**, 304–321 (2013). https://doi.org/10.1016/j.neuroimage.2012.12.062
23. Huber, P.J.: Robust estimation of a location parameter. Ann. Math. Stat. **35**, 73–101 (1964). https://doi.org/10.1214/aoms/1177703732
24. Akiba, T., Sano, S., Yanase, T., Ohta, T., Koyama, M.: Optuna: a next-generation hyperparameter optimization framework. In: 28th ACM SIGKDD Conference on Knowledge Discovery and Data Mining, pp. 2623–2631 (2019). https://doi.org/10.1145/3292500.3330701
25. Mccormick, M., Liu, X., Jomier, J., Marion, C., Ibanez, L.: ITK: enabling reproducible research and open science. Front. Neuroinform. **8**, 1–11 (2014). https://doi.org/10.3389/fninf.2014.00013
26. Wang, Z., Bovik, A.C., Sheikh, H.R., Simoncelli, E.P.: Image quality assessment: from error visibility to structural similarity. IEEE Trans. Image Process. **13**, 600–612 (2004). https://doi.org/10.1109/TIP.2003.819861
27. Vaserstein, L.N.: Markov processes over denumerable products of spaces, describing large systems of automata, Probl. Peredachi. Inf. **5**(3), 64–72 (1969)

Diagnostics, Aging, and Neurodegeneration

Non-parametric ODE-Based Disease Progression Model of Brain Biomarkers in Alzheimer's Disease

Matías Bossa[1(✉)], Abel Díaz Berenguer[1], and Hichem Sahli[1,2]

[1] Department of Electronics and Informatics (ETRO), Vrije Universiteit Brussel (VUB), 1050 Brussels, Belgium
{mnbossa,aberenguer,hsahli}@etrovub.be

[2] Interuniversity Microelectronics Centre (IMEC), 3001 Leuven, Belgium

Abstract. Data-driven disease progression models of Alzheimer's disease are important for clinical prediction model development, disease mechanism understanding and clinical trial design. Among them, dynamical models are particularly appealing because they are intrinsically interpretable. Most dynamical models proposed so far are consistent with a linear chain of events, inspired by the amyloid cascade hypothesis. However, it is now widely acknowledged that disease progression is not fully compatible with this conceptual model, at least in sporadic Alzheimer's disease, and more flexibility is needed to model the full spectrum of the disease. We propose a Bayesian model of the joint evolution of brain image-derived biomarkers based on explicitly modelling biomarkers' velocities as a function of their current value and other subject characteristics. The model includes a system of ordinary differential equations to describe the biomarkers' dynamics and sets a Gaussian process prior to the velocity field. We illustrate the model on amyloid PET SUVR and MRI-derived volumetric features from the ADNI study.

Keywords: Disease progression model · Alzheimer's disease (AD) · Magnetic resonance imaging (MRI) · Amyloid PET · Ordinary differential equations (ODE) · Gaussian process (GP)

1 Introduction

Alzheimer's disease (AD) is a growing health-economic worldwide issue, accounting for most cases of dementia [13]. Despite the great amount of effort devoted to AD prevention and drug development during the last three decades, the few pharmacological treatments available show a modest benefit. The study of the AD process is further hindered by the fact that dementia can be caused by multiple pathologies, and that AD often co-occurs with them [11], being age and genetic variations the main risk factors [10].

For more than two decades, the most widely accepted model of the pathophysiological process underlying AD was the so-called amyloid cascade hypothesis. This hypothesis states that the process starts with an abnormal accumulation of the β-amyloid (Aβ) peptide, triggering a chain of pathological events

A. Abdulkadir et al. (Eds.): MLCN 2022, LNCS 13596, pp. 95–103, 2022.
https://doi.org/10.1007/978-3-031-17899-3_10

in a predictable way. The corresponding model of biomarker dynamics states that the main AD biomarkers become abnormal in a temporally ordered manner [6,7]. However, large cohort studies showed that all possible combinations of biomarker abnormalities are frequently present in the cognitively normal population [8], evidencing that the amyloid cascade hypothesis is insufficient to explain the observed heterogeneity in sporadic AD [4,5].

A new conceptual model of AD was recently proposed [4], which posited a non-deterministic disease path. According to this model, Aβ and tau levels interact between them and with genetic and environmental factors to increase or reduce the risk of disease progression. These interactions would be responsible for the huge heterogeneity observed in biomarker trajectories and the discrepancies between observation and the amyloid cascade hypotheses.

Quantitative tools that estimate the biomarker dynamics are needed to shed light on the AD process and to build better clinical tools for diagnosis, prognosis and therapy efficacy assessment.

1.1 Disease Progression Models of Alzheimer's Disease

The first AD progression models describing long-term trajectories from short-term biomarker observations were based on Jack's model [7], *i.e.*, they assumed that all subjects follow the same disease progression pattern but with different onset times and at different speeds. Jedynak defined a disease progression score aimed at quantifying disease progression [9]. Subjects were temporally ordered according to this score and a parametric sigmoid-shaped curve was used to fit the progression of biomarkers. In [3] the authors proposed a semi-parametric model to determine the population mean of biomarker trajectories and the temporal order of subjects. A similar but more flexible model used Gaussian Process (GP) to model also the individual departures from the mean [12]. In general, all these models may suffer from identifiability issues when trained with short-term observations, because of the need to simultaneously estimate the disease onset times and the biomarker trajectories. Sometimes identifiability issues were mitigated using mixed-effect modelling to restrict the variance of the subject-level parameters.

The first dynamical model that relaxes the *unique trajectory* condition, allowing an arbitrary combination of variables as initial conditions, used a Riemannian framework to transport the mean trajectory to fit the subject's observations [16]. Contrary to the previous works, it is the initial value of the variables, and not the onset time, which was modelled as a random effect.

Finally, differential equation models parameterize biomarker velocities instead of biomarker trajectories and are therefore *implicit* models. Two works [2,15] tackle the problem of how to estimate long-term biomarker trajectories from short-term observations of a single biomarker. A recent work used a system of ordinary differential equations (ODEs) to simulate the effect of amyloid treatments on the disease course [1], being the first multivariate ODE-based AD progression model.

In this work, we propose a probabilistic AD progression model that uses a system of ODEs to describe biomarker dynamics. In our formulation, and similar to [16] and [1], all combinations of trajectory starting values are allowed. Another important common feature is that onset times are not model parameters, reducing the risk of non-identifiability. But contrary to all ODE-based approaches, we model the biomarker velocities non-parametrically, using GPs, which adds flexibility and imposes less inductive bias.

2 Methods

2.1 Definitions and Model Overview

We propose a Bayesian generative model to describe the trajectories of brain biomarkers throughout AD. Let $\mathbf{x}_s(t) = [x_{1,s}(t), x_{2,s}(t), \cdots, x_{L,s}(t)]$ be a set of L brain features and $\mathbf{y}_s(t) = [y_{1,s}(t), y_{2,s}(t), \cdots, y_{Q,s}(t)]$ a set of Q covariates for subject s at time t. The features $x_{l,s}(t)$ represent magnitudes associated with the disease status that evolve as the disease progress. For example, they could be brain atrophy, amyloid plaques or neurofibrillary tangles.In general, we can distinguish the aforementioned brain features, associated to the disease process, from the brain biomarkers extracted from MRI or PET images. However, we will consider one observable $\hat{x}_l$ per feature x_l and will refer to features and biomarkers interchangeably. The covariates $y_{q,s}(t)$ are assumed to have no observation error. They could be fixed over time (*e.g.* genetics), change in time according to a predefined or known pattern (*e.g.* age), or be controlled externally (*e.g.* treatments).

The link between a set of observed biomarkers $\hat{\mathbf{x}}$ and the feature vector $\mathbf{x}$ is specified by a likelihood function $\mathcal{L}(\hat{\mathbf{x}}|\mathbf{x}, \Theta)$, where Θ are model parameters. In our case, the likelihood functions will be independent Gaussian distributions. Let $\hat{x}_{l,s;i} \sim \mathcal{N}\left(x_{l,s}(t_{l,s;i}), \sigma_l^2\right)$ be the ith observation of biomarker l for subject s at observation time $t_{l,s;i}$. Note that the number observations and observation times may be different for each subject and each biomarker.

The main hypothesis in this work is that the state of features and covariates at a given time determines unequivocally the rate of progression,*i.e.* the expected rate of change of all the brain features. Specifically, trajectories should be a solution of the of the initial value problem defined by the system of ODEs

$$\frac{d\mathbf{x}_s(t)}{dt} = \mathbf{v}(\mathbf{z}_s(t)), \tag{1}$$

where $\mathbf{z}_s(t) = [\mathbf{x}_s(t), \mathbf{y}_s(t)]$, with initial condition

$$\mathbf{z}_s(0) = [\mathbf{x}_{0,s}, \mathbf{y}_s(0)]. \tag{2}$$

We propose to model each component l of the velocity field $\mathbf{v}(\cdot)$ using a GP prior

$$v_l \sim \mathcal{GP}(0, k_l(\cdot, \cdot)), \tag{3}$$

with the exponential kernel, $k_l(\mathbf{z}_m, \mathbf{z}_n) = k_{\alpha_l,\rho_l}(\mathbf{z}_m, \mathbf{z}_n) = \alpha_l^2 \exp(-\frac{(\mathbf{z}_m - \mathbf{z}_n)^2}{2\rho_l^2})$.

The problem is completely specified once we define priors for the model hyperparameters, *i.e.*, the observation variance σ_l^2, the kernel parameters α_l and ρ_l, and the subject-level parameters $\mathbf{x}_{0,s}$.

However, the velocity field $\mathbf{v}(\cdot)$ is a function-valued parameter which could be difficult to estimate and very hard or impossible to marginalize out given that it is involved in the ODE system (1).We propose the following approximation to transform Eq. (3) into a likelihood function and to model $\mathbf{v}(\cdot)$ implicitly, as it is usual in GP regression. Let $\hat{x}_{l,s;i}$ and $\hat{x}_{l,s;i+1}$ be two observations of the same biomarker l at two consecutive time points, $t_{l,s;i}$ and $t_{l,s;i+1}$, respectively. Assuming that the time difference $\Delta t_{l,s;i} = t_{l,s;i+1} - t_{l,s;i}$ is small with respect to the biomarker dynamics, we can approximate the velocity field using the observation differences.Dropping the indexes s and l for clarity, we have

$$\begin{aligned} x(t_{i+1}) &\simeq x(t_i) + v(\mathbf{z}(t_i))\Delta t_i \\ \hat{x}_{i+1} - \epsilon_{i+1} &\simeq \hat{x}_i - \epsilon_i + v(\mathbf{z}(t_i))\Delta t_i \\ \hat{v}_i = \frac{\hat{x}_{i+1} - \hat{x}_i}{\Delta t_i} &\simeq v(\mathbf{z}(t_i)) + \frac{\epsilon_{i+1} - \epsilon_i}{\Delta t_i}, \end{aligned} \tag{4}$$

where $\epsilon_i \sim \mathcal{N}(0, \sigma^2)$ is the observation error. Then $\hat{v}_i \sim \mathcal{N}(v(\mathbf{z}(t_i)), 2\sigma^2/\Delta t_i^2)$ and we can replace Eq. (3) with

$$\hat{v} \sim \mathcal{GP}(0, \hat{k}(\cdot,\cdot)), \tag{5}$$

where the new kernel $\hat{k}(\cdot,\cdot)$ is the same as $k(\cdot,\cdot)$ plus a noise term, and Eq. (5) represents a likelihood function because its l.h.s. is an observation. Note that the Gaussianity and independence of biomarker observations was critical to define the approximate velocities in Eq. (4).

2.2 Proposed Model

The complete set of parameter priors is given by

$$\begin{aligned} \sigma_l &\sim \mathcal{N}^+(0, \tau_{\sigma,l}) \\ \alpha_l &\sim \mathcal{N}^+(0, \tau_{\alpha,l}) \\ \rho_l &\sim \Gamma(5,5) \\ [x_{0,s}]_l &\sim \mathcal{U}(0,1), \end{aligned}$$

where $\mathcal{N}^+(\cdot,\cdot)$ is the half-Gaussian distribution and $\Gamma(5,5)$ is used as a weakly informative prior that penalizes extremely large and extremely small values of the length scale parameter ρ. Weakly informative priors for σ_l and α_l are determined by setting $\tau_{\sigma,l}$ and $\tau_{\alpha,l}$ equal to the mean subject-level standard deviation of observations and the variance of the estimated velocities $\hat{v}_{l,s;i}$, respectively.

For simplicity, each component of the subject-level parameters $\mathbf{x}_{0,s}$ is restricted to be in the unit segment after normalizing the biomarker values to fit in the unit hypercube. Note that they can be modelled as random effects if desired, which would provide better estimations in case that a biomarker is completely missing for a given subject.

The likelihood functions are

$$\begin{aligned} \hat{x}_{l,s;i} &\sim \mathcal{N}\left([\mathbf{x}_s(t_{l,s;i})]_l, \sigma_l^2\right) \\ \hat{\mathbf{v}}_l &\sim \mathcal{N}\left(\mathbf{0}, K_{\alpha_l,\rho_l}(\hat{Z}_l, \hat{Z}_l) + \operatorname{diag}\left(\mathbf{s}_{\sigma_l}^2\right)\right), \end{aligned} \tag{6}$$

where $\hat{\mathbf{v}}_l$ includes all consecutive observation differences from all subjects, *i.e.*,

$$\hat{\mathbf{v}}_l = \left[\frac{\hat{x}_{l,1;2} - \hat{x}_{l,1;1}}{\Delta t_{l,1;1}}, \frac{\hat{x}_{l,1;4} - \hat{x}_{l,1;3}}{\Delta t_{l,1;3}}, \ldots, \frac{\hat{x}_{l,2;2} - \hat{x}_{l,2;1}}{\Delta t_{l,2;1}}, \ldots\right]^T, \tag{7}$$

$\hat{Z}_l$ is a matrix with the corresponding features and covariates,

$$\hat{Z}_l = \begin{pmatrix} \hat{x}_{1,1;1} & \hat{x}_{2,1;1} & \cdots & y_{1,1;1} & y_{2,1;1} & \cdots \\ \hat{x}_{1,1;3} & \hat{x}_{2,1;3} & \cdots & y_{1,1;3} & y_{2,1;3} & \cdots \\ \vdots & \vdots & & \vdots & \vdots & \\ \hat{x}_{1,2;1} & \hat{x}_{2,2;1} & \cdots & y_{1,2;1} & y_{2,2;1} & \cdots \\ \vdots & \vdots & & \vdots & \vdots & \end{pmatrix} = \begin{pmatrix} (\hat{\mathbf{z}}_{1,1})^T \\ (\hat{\mathbf{z}}_{1,3})^T \\ \vdots \\ (\hat{\mathbf{z}}_{2,1})^T \\ \vdots \end{pmatrix},$$

the noise term is given by $\mathbf{s}_{\sigma_l} = \sqrt{2}\sigma_l[\Delta t_{l,1;1}^{-1}, \Delta t_{l,1;3}^{-1}, \ldots, \Delta t_{l,2;1}^{-1}, \ldots]$, and the matrix $K_{\alpha_l,\rho_l}(\hat{Z}_l, \hat{Z}_l)$ is obtained by applying the kernel to all combinations of rows in $\hat{Z}_l$.

To compute $\mathbf{x}_s(t_{l,s;i})$ in the likelihood term (Eq. (6)) we used forward Euler integration

$$\mathbf{x}_s(t_{l,s;i+1}) = \mathbf{x}_s(t_{l,s;i}) + \mathbf{v}(\mathbf{z}_s(t_{l,s;i}))\Delta t_{l,s;i}$$

with initial condition given by Eq. (2), where

$$[\mathbf{v}(\mathbf{z})]_l = \mathbf{k}_{\alpha_l,\rho_l}(\mathbf{z}, \hat{Z}_l)^T \left(K_{\alpha_l,\rho_l}(\hat{Z}_l, \hat{Z}_l) + \operatorname{diag}(\mathbf{s}_{\sigma_l}^2)\right)^{-1} \hat{\mathbf{v}}_l$$

Note that the trajectory of all biomarkers should be computed simultaneously, even when only a single component l is needed in Eq. (6). This doesn't represent any problem as far as the covariates $\mathbf{y}_s(t_{l,s;i})$ are available at all time points because the observations $\hat{x}_{l',s;i}$ are not used in this computations for $l' \neq l$. This implies that the biomarkers don't need to be acquired at the same time points. This is an important consideration for long longitudinal studies, such as ADNI, for which each imaging modality is scheduled at a different rate and the time gap between the first acquisition of two modalities differ between subjects.

However, the matrix $\hat{Z}_l$ should be complete for each component l. This implies that, at least at the first of the two time points used to compute the differences in Eq. (4), the complete set of observations is needed, and that the vectors $\hat{\mathbf{v}}_l$ in Eq. (7) may have different lengths for each biomarker l.

3 Experiments

3.1 Data

The model was fitted to the Alzheimer's Disease Neuroimaging Initiative (ADNI) database[1]. All subjects from the ADNI dataset having at least 4 valid AV45 (Florbetapir) PET scans and 4 MRI T1 scans were used to fit the model, resulting in a total of 198 participants (88 Cognitively Normal and 110 with Mild Cognitive Impairment), and 874 PET and 1225 MRI measurements.

Three features were selected: mean AV45-PET SURV (average PET signal in cortical grey matter normalized by whole cerebellum), and the ratios of hippocampal and ventricular volume to intracranial volume (ICV)[2]. Covariates included age and the presence of a copy of the E4 allele of the apolipoprotein-E (APOE) gene. The covariate vector $\mathbf{y}$ had only 1 dimension (age) because velocities fields for APOE E4 carriers and non-carriers were kept apart and estimated separately, sharing only the hyperparameters.

3.2 Results

The posterior distributions of the model parameters were obtained with Markov chain Monte Carlo (MCMC) sampling using Stan software [17]. To explore the model predictive performance we did a leave-one-site-out experiment consisting in removing all subjects from a given hospital, except for a few observations used for predicting the rest of the biomarker trajectory, but not for velocity estimation.All the observations from left-out sites within a 5-year interval centred at the PET-MRI time overlap were used to estimate the rest of the trajectory. This interval was defined in order that all subjects to have at least one PET observation. As AV45 started to be acquired after MRI, there are few observations after this period. Therefore, prediction time in the past is larger than in the future.

Figure 1 shows predicted trajectories along with observations for a subset of subjects. Specifically, we selected the subject with most data not shown to the model from each site. Then, the 8 subjects with the longest unobserved trajectories from the APOE E4 non-carrier group were selected.

Apart from prediction, the model allows to test an endless amount of hypotheses, such as the mean difference in a given biomarker rate of change between two given sub-populations. For the sake of illustration, we have focused on the hippocampal rate of change, shown in Fig. 2. The top row panels show a representation of the velocity field in the MRI plane (hippocampal and ventricular volumes) and the bottom row panels show the rate of change of hippocampal volume for different conditions. The most prominent pattern is that APOE E4 carriers present higher rates of hippocampal atrophy than E4 non-carriers. These dynamics are only mildly modulated by brain amyloid levels, as can be observed in the bottom row panels. The strong influence of genetic factors in AD dynamics and the importance of considering their effect in AD progression modelling was recently highlighted in [4].

---

[1] adni.loni.usc.edu.

[2] The MRI volumes were computed using FreeSurfer (5.1).

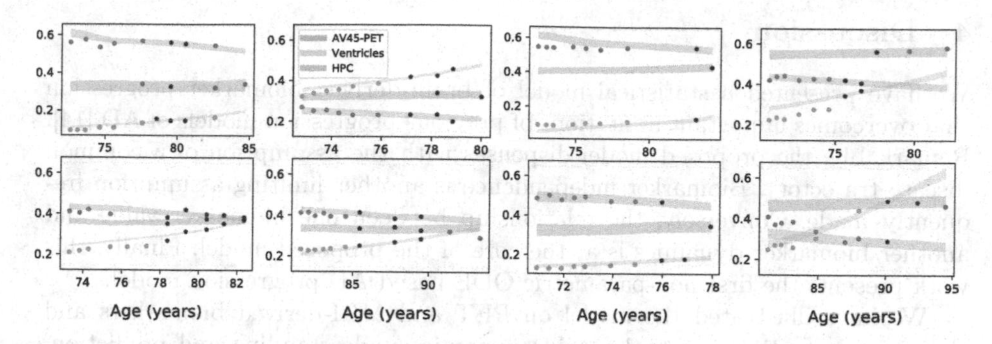

Fig. 1. Normalized biomarker trajectories for selected individuals. Blue: AV45-PET mean SUVR/3; Orange: Ventricular volume/ICV ×10; Green: Hippocampal (HPC) volume/ICV ×100. Dots correspond to observations, black stars denote observations used for prediction while the rest of the observations were hidden during the model inference. Shaded areas represent the 90% highest density interval of the prediction posterior probability. (Color figure online)

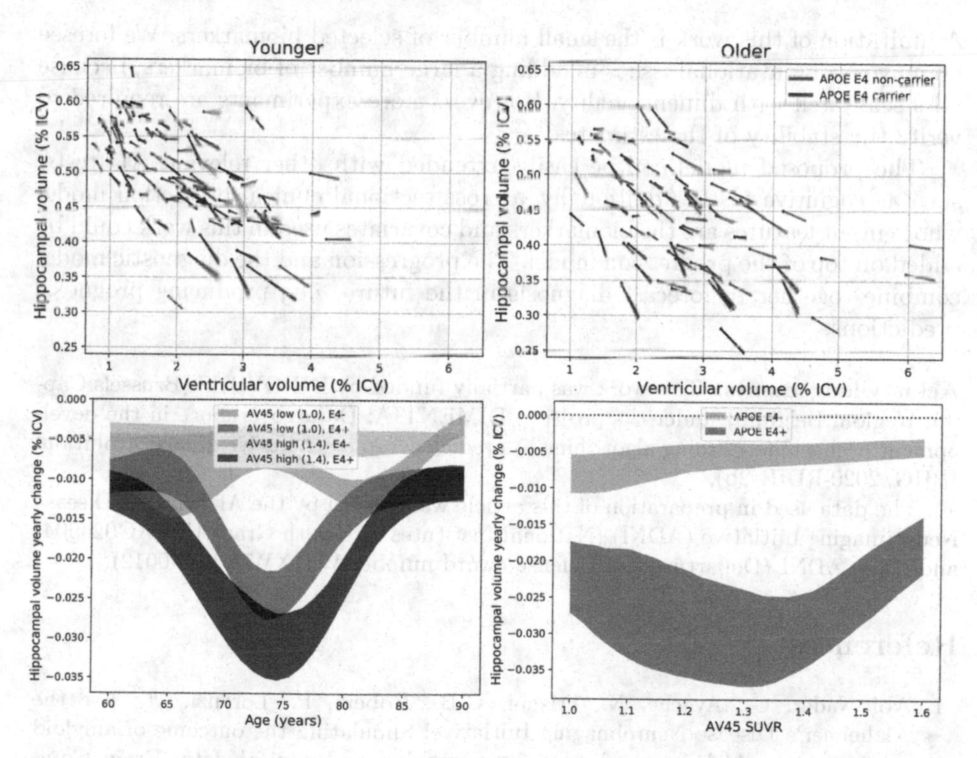

Fig. 2. Top: Representation of velocity fields. Dots correspond to the mean estimated initial values and lines represent velocity (two years of evolution) and their uncertainty (lines are drawn from the posterior distribution). **Bottom**: Hippocampal volume rate of change among subjects with mild neurodegeneration for different conditions of age, APOE and amyloid PET. Shaded areas represent 90% highest density interval of the posterior probability.

4 Discussion

We have presented a statistical model of brain-derived biomarker progression that overcomes important limitations of previous progression models of AD [14]. Remarkably, the proposed model dispenses with the assumption of a common disease trajectory. Biomarker independence is another limiting assumption frequently made. Conversely, the relationship between one biomarker value and another biomarker dynamics is at the core of the proposed model. Finally, this work presents the first non-parametric ODE-based AD progression model.

We have illustrated the model on PET and MRI-derived biomarkers and shown its potential as a tool for AD dynamics understanding and prediction. We showed the feasibility of full Bayesian posterior inference using MCMC in a moderate-sized dataset. The only multivariate ODE-based model of AD progression we are aware of used a variational approximation to estimate posteriors [1].

4.1 Limitations and Future Directions

A limitation of this work is the small number of selected biomarkers. We foresee no mayor computational issues in adding a large number of biomarkers, because GPs scale well with dimensionality. However, more experiments are required to verify the stability of the estimates.

The proposed model can be easily extended with other relevant AD tests, such as cognitive tests. Additionally, a cross-sectional clinical prediction model whose input features are the biomarkers and covariates used in this work could be added on top of the progression model. The progression and the diagnostic model combine together to forecast diagnosis in the future, *i.e.*, producing prognosis predictions.

Acknowledgements. This work was partially funded by INNOVIRIS (Brussels Capital Region, Belgium) under the project: 'DIMENTIA: Data governance in the development of machine learning algorithms to predict neurodegenerative disease evolution' (BHG/2020-RDIR-2b).

The data used in preparation of this article was funded by the Alzheimer's Disease Neuroimaging Initiative (ADNI) (National Institutes of Health Grant U01 AG024904) and DOD ADNI (Department of Defense award number W81XWH-12-2-0012).

References

1. Abi Nader, C., Ayache, N., Frisoni, G.B., Robert, P., Lorenzi, M.: For the Alzheimer's Disease Neuroimaging Initiative: Simulating the outcome of amyloid treatments in Alzheimer's disease from imaging and clinical data. Brain Commun. **3**(2), 1–17 (2021). https://doi.org/10.1093/braincomms/fcab091. https://academic.oup.com/braincomms/article-pdf/3/2/fcab091/38443216/fcab091.pdf
2. Budgeon, C., Murray, K., Turlach, B., Baker, S., Villemagne, V., Burnham, S.: For the Alzheimer's Disease neuroimaging initiative: constructing longitudinal disease progression curves using sparse, short-term individual data with an application to Alzheimer's disease. Stat. Med. **36**(17), 2720–2734 (2017)

3. Donohue, M.C., et al.: Estimating long-term multivariate progression from short-term data. Alzheimer's Dementia **10**(5, Supplement), S400–S410 (2014)
4. Frisoni, G.B., et al.: The probabilistic model of Alzheimer disease: the amyloid hypothesis revised. Nat. Rev. Neurosci. **23**(1), 53–66 (2022)
5. Herrup, K.: The case for rejecting the amyloid cascade hypothesis. Nat. Neurosci. **18**(6), 794–799 (2015)
6. Jack, C.R., et al.: Tracking pathophysiological processes in Alzheimer's disease: an updated hypothetical model of dynamic biomarkers. Lancet Neurol. **12**(2), 207–216 (2013)
7. Jack, C.R., et al.: Hypothetical model of dynamic biomarkers of the Alzheimer's pathological cascade. Lancet Neurol. **9**(1), 119–128 (2010)
8. Jack, C.R., et al.: Age-specific and sex-specific prevalence of cerebral β-amyloidosis, tauopathy, and neurodegeneration in cognitively unimpaired individuals aged 50–95 years: a cross-sectional study. Lancet Neurol. **16**(6), 435–444 (2017)
9. Jedynak, B.M., et al.: A computational neurodegenerative disease progression score: Method and results with the Alzheimer's disease neuroimaging initiative cohort. Neuroimage **63**(3), 1478–1486 (2012)
10. Knopman, D.S., et al.: Alzheimer Disease. Nat. Rev. Dis. Primers. **7**(1), 33 (2021)
11. Kovacs, G.G., et al.: Non-Alzheimer neurodegenerative pathologies and their combinations are more frequent than commonly believed in the elderly brain: a community-based autopsy series. Acta Neuropathol. **126**(3), 365–384 (2013)
12. Lorenzi, M., Filippone, M., Frisoni, G.B., Alexander, D.C., Ourselin, S.: Probabilistic disease progression modeling to characterize diagnostic uncertainty: application to staging and prediction in Alzheimer's disease. Neuroimage **190**, 56–68 (2019)
13. Nichols, E., et al.: Estimation of the global prevalence of dementia in 2019 and forecasted prevalence in 2050: an analysis for the global burden of disease study 2019. Lancet Public Health **7**(2), e105–e125 (2022)
14. Oxtoby, N.P., Alexander, D.C.: For the EuroPOND consortium: Imaging plus x: multimodal models of neurodegenerative disease. Curr. Opinion Neurol. **30**(4), 371–379 (2017)
15. Oxtoby, N.P., et al.: Learning imaging biomarker trajectories from noisy Alzheimer's Disease data using a Bayesian multilevel model. In: Cardoso, M.J., Simpson, I., Arbel, T., Precup, D., Ribbens, A. (eds.) BAMBI 2014. LNCS, vol. 8677, pp. 85–94. Springer, Cham (2014). https://doi.org/10.1007/978-3-319-12289-2_8
16. Schiratti, J.B., Allassonnière, S., Colliot, O., Durrleman, S.: A bayesian mixed-effects model to learn trajectories of changes from repeated manifold-valued observations. J. Mach. Learn. Res. **18**(133), 1–33 (2017)
17. Stan Development Team: Stan Modeling Language Users Guide and Reference Manual, Version 2.29 (2019). https://mc-stan.org

Lifestyle Factors That Promote Brain Structural Resilience in Individuals with Genetic Risk Factors for Dementia

Elizabeth Haddad[1], Shayan Javid[1], Nikhil Dhinagar[1], Alyssa H. Zhu[1], Pradeep Lam[1], Iyad Ba Gari[1], Arpana Gupta[2], Paul M. Thompson[1], Talia M. Nir[1], and Neda Jahanshad[1](✉)

[1] Imaging Genetics Center, Mark and Mary Stevens Neuroimaging and Informatics Institute, Keck School of Medicine, University of Southern California, Los Angeles, CA, USA
njahansh@usc.edu
[2] David Geffen School of Medicine, University of California, Los Angeles, Los Angeles, CA, USA

Abstract. Structural brain changes are commonly detectable on MRI before the progressive loss of cognitive function that occurs in individuals with Alzheimer's disease and related dementias (ADRD). Some proportion of ADRD risk may be modifiable through lifestyle. Certain lifestyle factors may be associated with slower brain atrophy rates, even for individuals at high genetic risk for dementia. Here, we evaluated 44,100 T1-weighted brain MRIs and detailed lifestyle reports from UK Biobank participants who had one or more genetic risk factors for ADRD, including family history of dementia, or one or two ApoE4 risk alleles. In this cross-sectional dataset, we use a machine-learning based metric of age predicted from cross-sectional brain MRIs - or 'brain age' - which when compared to the participant's chronological age, may be considered a proxy for abnormal brain aging and degree of atrophy. We used a 3D convolutional neural network trained on T1w brain MRIs to identify the subset of genetically high-risk individuals with a substantially lower brain age than chronological age, which we interpret as resilient to neurodegeneration. We used association rule learning to identify sets of lifestyle factors that were frequently associated with brain-age resiliency. Never or rarely adding salt to food was consistently associated with resiliency. Sex-stratified analyses showed that anthropometry measures and alcohol consumption contribute differently to male vs female resilience. These findings may shed light on distinctive risk profile modifications that can be made to mitigate accelerated aging and risk for ADRD.

Keywords: Alzheimer's disease · Brain age · Lifestyle factors

1 Introduction

Late-onset Alzheimer's disease (LOAD) and related dementias, which are considered to be those that occur after the age of 65, are complex disorders, driven by a combination and wide range of genetic, environment, and gene-environment interactions [1].

A. Abdulkadir et al. (Eds.): MLCN 2022, LNCS 13596, pp. 104–114, 2022.
https://doi.org/10.1007/978-3-031-17899-3_11

Recently, publicly available big data initiatives and biobanks have taken a more holistic approach and collect a wide range of data to help understand sources of risk factors for diseases, including ADRDs. Machine learning techniques provide computationally efficient and effective means of data reduction and analyses that can promote our understanding of neurodegenerative processes from the vast set of data features now available to researchers; brain imaging and genetic data alone can contain trillions of univariate combinations of tests [2]. Here, we use machine learning approaches to discover lifestyle factors that may contribute to brain resilience in individuals at heightened genetic risk for ADRDs.

People with a family history of dementia have twice the risk of being affected themselves than individuals in the general population [3]. Apolipoprotein (ApoE) E4 (e4) on chromosome 19 has consistently been shown to be the commonly-occurring genetic factor most strongly associated with LOAD [4]: one copy of e4 poses an approximate 3–5-fold increase in lifetime risk, whereas two copies pose an 8–20-fold increase [3, 5], depending on a person's ancestry. Modifiable risk factors also contribute to risk for LOAD, and together are associated with approximately 40% of the population attributable risk for dementia worldwide [6, 7]. These include less education, hearing loss, traumatic brain injury, hypertension, excessive alcohol consumption, obesity, smoking, depression, social isolation, physical activity, air pollution, and diabetes. Other factors that have been associated with Alzheimer's pathology and cognitive decline include poor sleep and dietary factors, respectively [6]. As LOAD is characterized by brain structural changes, such as cortical and hippocampal atrophy [8], studying the impact of modifiable factors on these changes may inform preventative, high yield recommendations for those at risk.

Typically, longitudinal data is required to infer rates of brain structural changes that occur as a result of age or disease. Over the last decade, a surrogate cross-sectional marker for measuring "accelerated" or "slowed" brain aging, more commonly referred to as "brain age", has been developed [9]. Although not without limitations [10], brain age is derived from machine learning models, which often rely on regression techniques that model brain features as independent variables and chronological age as the dependent variable in a neurologically healthy "training" set. These models are then applied to an independent or "test" sample and the difference between an individual's predicted age and chronological age, known as the "brain age gap" or residual, is used as an aging biomarker. Brain age has been used to study associations of risk factors with accelerated aging, predict diagnostic outcomes, and examine differences between clinical populations. Recently, with the use of deep learning techniques such as convolutional neural networks (CNNs), researchers can now predict brain age with less overall error, and more precise age estimates, directly from the full 3D MRI scan [11].

Most studies to date have used brain age in older adults to model accelerated aging, but investigating associations with "slowed" brain aging is also important to better understand factors that can mitigate accelerated brain aging. Continuous measures of brain age have shown protective associations with physical exercise, education, and combined lifestyle scores [12–15] but no study to date has evaluated which combination of these and other modifiable factors most consistently promotes brain resilience. The UK Biobank is a large, densely phenotyped epidemiological study that has collected health information from half a million UK participants, approximately 50,000 with neuroimaging [16]. This has allowed researchers to comprehensively examine how genetic, sociodemographic, physical, and health outcomes may influence brain structure and function. In such high dimensional datasets, data mining algorithms, including association rule learning (ARL) can be used to discover associations between sets of features and a particular outcome [17]. By combining ARL with measures of brain aging in deeply phenotyped datasets, we can begin to make inferences that incorporate the multifactorial nature of complex neurodegenerative processes and dementia risk.

Using data from the UK Biobank, we identified high LOAD risk individuals – who are over age 65 and either ApoE4 carriers or have a family history of dementia – who are resilient to brain aging, characterized by a brain age younger than their chronological age; using ARL, we then examined which sets of lifestyle factors were most often associated with this resiliency.

2 Methods

2.1 "Brain Age"

44,100 T1-weighted (T1w) brain MRI scans from the UK Biobank (UKB) [16] were preprocessed using a sequence of standard neuroimaging processing steps including N4 bias field correction for non-parametric intensity normalization, skull stripping, and linear registration with 6 degrees of freedom to a template. The preprocessed T1w scans from UKB were partitioned into non-overlapping training and test splits using a 5-fold cross-validation approach. We used 20% of each training split as a validation set. Our model architecture for brain age prediction is based on the 3D CNN originally proposed by [18]. The model consists of a sequence of five convolutional blocks with a [Conv-BatchNorm-Activation-MaxPooling] structure as the backbone, followed by two fully connected layers. The architecture is relatively lightweight for a 3D-CNN, with around two million parameters. We trained the CNN end-to-end for 30 epochs with a batch size of 5 using the Adam optimizer [19] and learning rate of 0.00005. We extracted the brain age measures for all 44,100 subjects with available T1w MRI at the time of writing. We corrected for predicted age's dependence on age to avoid artificially inflating performance metrics [10]. In brief, this involves scaling the predicted age by the slope and intercept from a regression of predicted age on chronological age [20, 21].

2.2 Subject Selection

UKB participants over age 65 with e3e4 or e4e4 genotype or a family history dementia (maternal, paternal, or sibling) were included (4520M, 4493F). The mean absolute error (MAE) of our bias-corrected brain age model was calculated and any subject having a brain age lower than their true age by a value larger than the MAE was considered resilient.

2.3 Lifestyle Factors

Lifestyle factors were derived and binarized from lifestyle and environment, anthropometry, and summary diagnostic categories measures, documented at the time of imaging. Factors and their respective datafield IDs are provided in Table 1. Body mass index classifications were categorized under CDC guidelines. Diet quality was modeled using the same coding scheme as in [22] and [23]. Briefly, an overall diet quality score was computed and binarized as adequate if the score was >50 out of a total of 100. Individual diet components were also included and corresponded to adequate consumption of fruits, vegetables, whole grains, fish, dairy, vegetable oil, refined grains, processed meat, unprocessed meat, and sugary food/drink intake.

Table 1. Lifestyle factors and respective UK Biobank data field IDs. American Heart Association (AHA) guidelines for weekly ideal ($\geq$150 min/week moderate or $\geq$ 75 min/wk vigorous or 150 min/week mixed), intermediate (1–149 min/week moderate or 1–74 min/week vigorous or 1–149 min/week mixed), and poor (not performing any moderate or vigorous activity) physical activity. Supplementation was categorized into any vitamins/minerals or fish oil intake. Salt added to food and variation in diet included the response of "never or rarely". Smoking status included never having smoked, previously smoked, and currently smokes. Alcohol frequency was categorized as infrequent (1–3 times a month, special occasions only, or never), occasional (1–2 a week or 3–4 times a week), and frequent (self-report of daily/almost daily and ICD conditions F10, G312, G621, I426, K292, K70, K860, T510). Social support/contact variables included attending any type of leisure/social group events, having family/friend visits twice a week or more, and being able to confide in someone almost daily.

Lifestyle factor	Features (data field ID)
Physical activity/Body composition	AHA Physical Activity (884, 904, 894, 914); Waist to Hip Ratio (48, 49); Body Mass Index (BMI) (23104)
Sleep	Sleep 7–9 h a Night (1160); Job Involves Night Shift Work (3426); Daytime Dozing/Sleeping (1220)
Diet/Supplements	Diet Quality Score and Components (based on Said et al., 2018; Zuang et al., 2021) (1309, 1319, 1289, 1299, 1438, 1448, 1458, 1468, 1329, 1339, 1408, 1418, 1428, 2654, 1349, 3680, 1359, 1369, 1379, 1389, 3680, 6144); Fish Oil Supplementation (20084); Vitamin/Mineral Supplementation (20084); Salt Added to Food (1478); Variation in Diet (1548)
Education	College/University (6138)
Smoking	Smoking Status (20116)
Alcohol	Alcohol Intake Frequency (1558/ICD)
Social contact/Support	Attending Leisure/Social Group Events (6160); Frequency of Friends/Family Visits (1031); Able to Confide in Someone (2110)

2.4 Association Rule Learning

ARL using the *mlxtend* library in Python [24] was used to characterize sets of lifestyle factors that co-occur with resilience across the older population, and after stratifying for sex.

Problem Formulation. Let $I = \{i_i, \ i_2, ..., i_m\}$ be a set of binary attributes (items), here, being the resilience classification, and the set of binary lifestyle factors described in Table 1. Let $T = \{t_1, t_2, ..., t_N\}$ be the set of subjects. Each subject is represented as a binary vector where $t_j[i_k] = 1$ if the subject j, t_j, has that particular feature i_k, and $t_j[i_k] = 0$ otherwise. An itemset, X, is defined by a collection of zero or more items. In this context, an association rule is an implication of the form $X \Rightarrow i_k$, where X is a set of lifestyle factors in I and our consequent of interest, i_k, is the item "resilience", with $|X \cap \{i_k\}| = 0$. ARL is decomposed into two subproblems:

1. Frequent itemset generation, where the objective is to find itemsets that are above some minimum support and
2. Rule generation, where the objective is to generate rules with high lift values (defined below) from the frequent itemsets generated in step 1.

Frequent Itemset Generation. For the first step of the association rule generation, the *Apriori* principle states that, if an itemset is frequent then all of its subset itemsets are also frequent [25]. The metric that can be used to generate frequent itemsets is support. Support is defined as the fraction of subjects (i.e., transactions) in T that contain that itemset. That is

$$support(Z) = \frac{\sigma(Z)}{N} \tag{1}$$

where Z is an itemset, N is the total number of subjects, and $\sigma(Z)$ is the support count of Z defined as

$$\sigma(Z) = |\{t_i | Z \subseteq t_i, t_i \in T\}| \tag{2}$$

where $|\cdot|$ represents the cardinality of the set. We generated all the frequent itemsets with a minimum support threshold of 0.01. The minimum support threshold was chosen based on the computational power and memory available to us, we have chosen a very small value (1%) to incorporate as many itemsets as possible for our association rule generation.

Rule Generation. For each frequent itemset f_k generated, we would generate the association rules that have "resilient" as their consequent. To measure the strength of a rule we used the lift value:

$$lift(X \Rightarrow i_k) = \frac{support(X \cap \{i_k\})}{support(X) \times support(\{i_k\})} \tag{3}$$

A lift value greater than 1 implies that the degree of association between the antecedent and consequent is higher than in a situation where the antecedent and consequent are independent. We generated all the association rules with a minimum lift value of 1.2.

3 Results

3.1 Brain Age

The trained models were evaluated on the test set using the root mean squared error (RMSE), mean absolute error (MAE), and Pearson's r between the true age and predicted brain age as seen in Table 2 below. The 3D CNN achieved an average MAE of 2.91(0.05) across the 5 splits. Following age bias correction, the MAE was 3.21.

Table 2. Summary of the test performance of the 3D CNN on UKB for brain age prediction in the full sample of N = 44,100. RMSE: root mean squared error; MAE: mean absolute error.

	Split 1	Split 2	Split 3	Split 4	Split 5
RMSE	3.747	3.676	3.712	3.604	3.584
MAE	2.972	2.905	2.958	2.864	2.84
Pearson's r	0.883	0.878	0.888	0.885	0.887

3.2 Resiliency

1439/9013 (16%) of subjects over age 65 years with genetic risk factors for ADRD were considered to be resilient, that is, having a brain age value > 3.21 years younger than their chronological age. 760/4520 (17%) of females and 679/4493 (15%) of males were resilient. Example subjects are featured in Fig. 1.

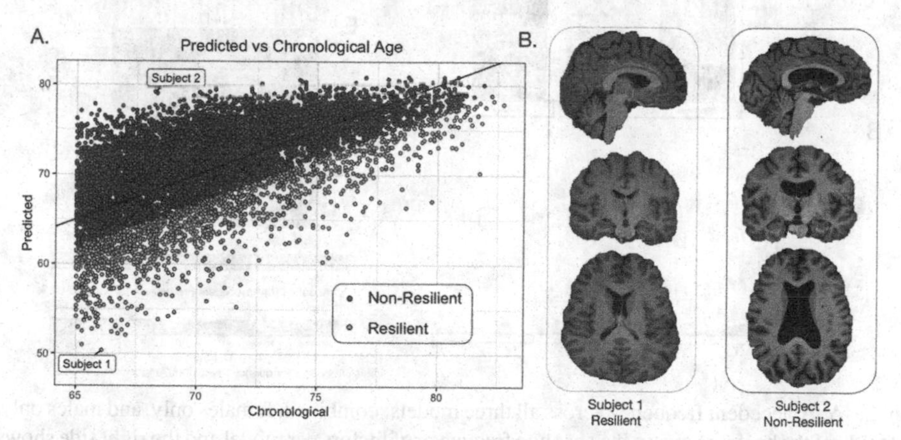

Fig. 1. A. Predicted (y-axis) age plotted against chronological (x-axis) age **B.** Brain structural T1-weighted MRIs are shown for a resilient case (subject 1; actual age: 66; predicted age: 50) and a non-resilient case (subject 2; actual age: 68; predicted age: 79). Note qualitative differences in ventricular size are visible, even though subjects are close in chronological age.

3.3 Association Rule Learning

A total of 7,076 sets with a minimum support of 0.01 and lift value of at least 1.2 co-occurred with resiliency in the combined set, 34,106 for females, and 5,883 for males. The top antecedent set for each model is shown in Table 3. Frequencies of factors are visualized in Fig. 2A for all three models. Corresponding lift values are shown in Fig. 2B.

The most frequent factor that appeared in antecedent sets for combined (69.7%) and female (61.6%), was "never or rarely adding salt to food", but also occurred with high frequency in males (49.5%). The most frequent factor in males was having an adequate

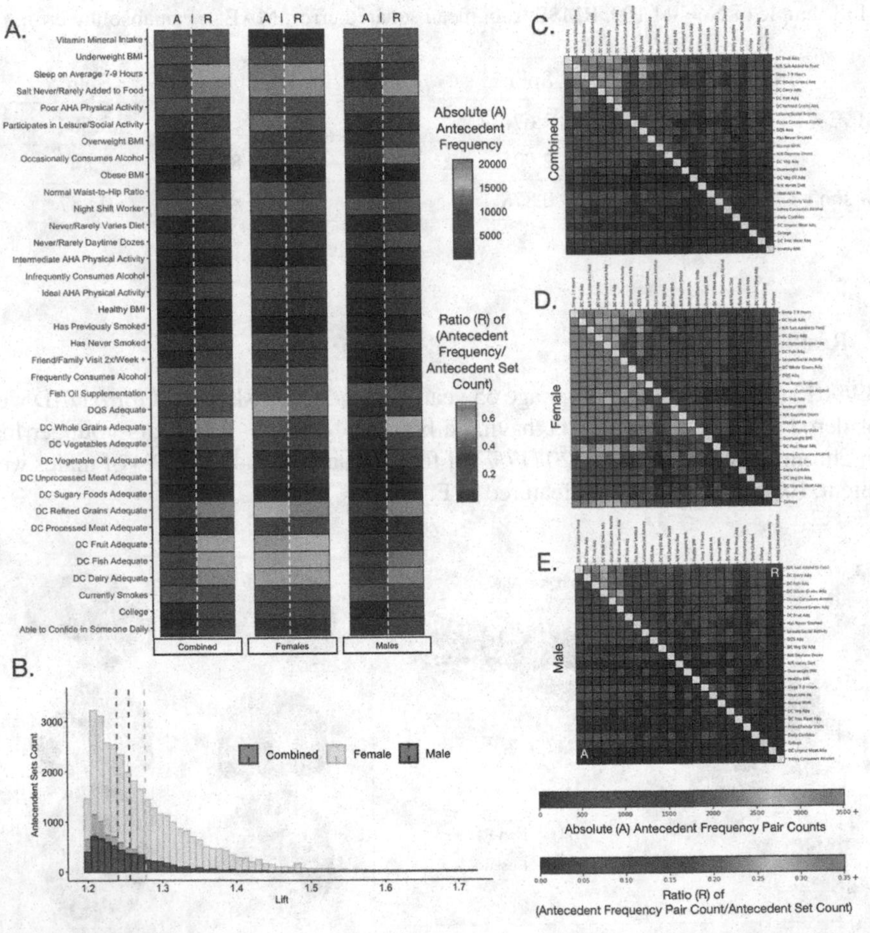

Fig. 2. **A.** Antecedent frequency across all three models: combined, females only, and males only. Left side of the heatmap shows the absolute frequency of factors per model and the right side shows a ratio of the absolute count over the total number of predictive sets. **B.** Corresponding lift values and means (dotted line) for each respective model. Highest pair frequencies of lifestyle factors in antecedent sets in the **C.** combined, **D.** female, and **E.** male models. Lower triangles are absolute frequencies truncated at 3500 and upper triangles indicate ratio of absolute pair counts over the total number of predictive sets truncated at 0.35.

intake of whole grains (54.1%) which also occurred with high frequency in the combined set (48.3%) and in the female set (36.3%) although to a lesser extent. Having an adequate diet quality score was in 39% of the combined sets, 45.4% of the female sets, and 28.4% of the male sets. Sleeping on average 7–9 h per night was in 60.4% of the female, 57.4% of the combined, and 31.4% of the male sets. Participating in leisure or social activities appeared at a relatively similar rate among combined (35.4%), female (38.8%), and male (33.5%) models.

Factors that were in less than 1% of predictive sets in any of the combined, male and female models included: obese or underweight BMI; frequent consumption of alcohol; night shift work; current smoking status; having poor physical activity; and supplementation of vitamins or fish oil.

Table 3. Top antecedent set with resiliency as a consequent based on lift from the combined (**C**), female (**F**), and male (**M**) models.

	Antecedents (A), resiliency as consequent (C)	A support	C support	Support	Lift
C	infrequent alcohol, never/rarely varies diet, adequate fruit, never smoked	0.043	0.160	0.010	1.46
F	never/rarely daytime dozes, never/rarely varies diet, adequate fish, adequate dairy, friend/family visits, infrequent alcohol	0.035	0.168	0.010	1.72
M	never/rarely varies diet, adequate dairy, college, adequate processed meat, never smoke	0.042	0.151	0.010	1.59

4 Discussion

As the prevalence of ADRDs continues to rise, understanding which lifestyle modifications can mitigate brain aging risk is becoming increasingly important. Here, in those at elevated risk, we identified subjects resilient to brain aging and examined the lifestyle factors most associated with this resiliency. The self reporting of "never or rarely adding salt to food" was frequently associated with resilience across all of our models tested. Dietary salt is associated with hypertension, a known vascular risk factor for dementia. However, irrespective of hypertension, independent associations between a high salt diet and increased dementia risk have been shown [26–28]. We also found support for other modifiable risk factors such as diet quality, social contact, the absence of smoking, physical activity, and sleep duration, as we detected associations with these factors and resilience.

As studies continue to show differential risk profiles across the sexes [7], we further extend our work by building sex-specific models. Not only is the incidence of AD greater in females, but carrying an e4 allele has a stronger effect in females compared to males [29–31]. Although e4 status and sex are non-modifiable, it is important to investigate how they differentially interact with factors that *are* modifiable [32]. Interestingly, we found nearly 5 times the number of predictive sets in females compared to the combined model and the males only model.

Our results are largely based on self-reported data from predetermined questionnaires, which may be confounded by reporting and response biases. Moreover, ARL requires binary variables which reduces the granularity of some of the features – a factor that may play an important role in predicting resiliency. We also note limitations associated with the use of brain age gap, as a type of residual method, given the lack of ground truth data needed to assess its validity [33]. Our brain age method also lacks spatial information regarding anatomical changes that occur as a result of brain aging. Future work will address these limitations by modeling specific brain features as outcome variables in order to spatially map which features most consistently contribute to brain aging resilience. Continuous lifestyle factors will be modeled as predictors using methods that are robust to potential hidden confounders [34].

Nonetheless, with the ability to map the discrepancies between chronological age and what may be more functionally important – biological age – we can begin to understand which modifiable factors are more or less beneficial to brain health. More importantly, by building stratified models, we can shed light on differential risk profiles that may inform respective populations on the most effective actions to take to mitigate accelerated brain aging and dementia risk. Future work will continue to tease apart potential differential effects by separately modeling e4 carriage and familial history of dementia as they relate to resilience, and modeling sex-specific brain age resilience.

Acknowledgments. This work was supported in part by: R01AG059874, U01AG068057, P41EB05922. This research has been conducted using the UK Biobank Resource under Application Number '11559'.

References

1. Blennow, K., de Leon, M.J., Zetterberg, H.: Alzheimer's disease. Lancet (2006)
2. Medland, S.E., Jahanshad, N., Neale, B.M., Thompson, P.M.: Whole-genome analyses of whole-brain data: working within an expanded search space. Nat. Neurosci. **17**, 791–800 (2014)
3. Loy, C.T., Schofield, P.R., Turner, A.M., Kwok, J.B.J.: Genetics of dementia. Lancet **383**, 828–840 (2014)
4. Lambert, J.C., Ibrahim-Verbaas, C.A., Harold, D., Naj, A.C., Sims, R., et al.: Meta-analysis of 74,046 individuals identifies 11 new susceptibility loci for Alzheimer's disease. Nat. Genet. **45**, 1452–1458 (2013)
5. Strittmatter, W.J.: Medicine. Old drug, new hope for Alzheimer's disease. Science **335**(6075), 1447–1448 (2012)
6. Livingston, G., et al.: Dementia prevention, intervention, and care: 2020 report of the Lancet Commission. Lancet **396**, 413–446 (2020)
7. Nianogo, R.A., Rosenwohl-Mack, A., Yaffe, K., Carrasco, A., Hoffmann, C.M., Barnes, D.E.: Risk factors associated with Alzheimer disease and related dementias by sex and race and ethnicity in the US. JAMA Neurol. **79**, 584–591 (2022)
8. Jack, C.R., Jr.: Alliance for aging research AD biomarkers work group: structural MRI. Neurobiol. Aging. **32**(Suppl 1), S48-57 (2011)

9. Cole, J.H., Franke, K.: Predicting age using neuroimaging: innovative brain ageing biomarkers. Trends Neurosci. **40**, 681–690 (2017)
10. Butler, E.R., Chen, A., Ramadan, R., Le, T.T., Ruparel, K., et al.: Pitfalls in brain age analyses. Hum. Brain Mapp. **42**, 4092–4101 (2021)
11. Cole, J.H., Poudel, R.P.K., Tsagkrasoulis, D., Caan, M.W.A., Steves, C., et al.: Predicting brain age with deep learning from raw imaging data results in a reliable and heritable biomarker. Neuroimage **163**, 115–124 (2017)
12. Steffener, J., Habeck, C., O'Shea, D., Razlighi, Q., Bherer, L., Stern, Y.: Differences between chronological and brain age are related to education and self-reported physical activity. Neurobiol. Aging. **40**, 138–144 (2016)
13. Smith, S.M., Vidaurre, D., Alfaro-Almagro, F., Nichols, T.E., Miller, K.L.: Estimation of brain age delta from brain imaging. Neuroimage **200**, 528–539 (2019)
14. Dunås, T., Wåhlin, A., Nyberg, L., Boraxbekk, C.-J.: Multimodal image analysis of apparent brain age identifies physical fitness as predictor of brain maintenance. Cereb. Cortex. **31**, 3393–3407 (2021)
15. Bittner, N., et al.: When your brain looks older than expected: combined lifestyle risk and BrainAGE. Brain Struct. Funct. **226**(3), 621–645 (2021). https://doi.org/10.1007/s00429-020-02184-6
16. Miller, K.L., Alfaro-Almagro, F., Bangerter, N.K., Thomas, D.L., Yacoub, E., et al.: Multimodal population brain imaging in the UK Biobank prospective epidemiological study. Nat. Neurosci. **19**, 1523–1536 (2016)
17. Agrawal, R., Imieliński, T., Swami, A.: Mining association rules between sets of items in large databases. In: Proceedings of the 1993 ACM SIGMOD International Conference on Management of Data - SIGMOD 1993. ACM Press, New York (1993)
18. Peng, H., Gong, W., Beckmann, C.F., Vedaldi, A., Smith, S.M.: Accurate brain age prediction with lightweight deep neural networks. Med. Image Anal. **68**, 101871 (2021)
19. Kingma, D.P., Ba, J.: Adam: a method for stochastic optimization, http://arxiv.org/abs/1412.6980 (2014)
20. Cole, J.H., Ritchie, S.J., Bastin, M.E., Valdés Hernández, M.C., Muñoz Maniega, S., et al.: Brain age predicts mortality. Mol. Psychiatry. **23**, 1385–1392 (2018)
21. de Lange, A.-M.G., Cole, J.H.: Commentary: correction procedures in brain-age prediction. Neuroimage Clin. **26** (2020)
22. Said, M.A., Verweij, N., van der Harst, P.: Associations of combined genetic and lifestyle risks with incident cardiovascular disease and diabetes in the UK Biobank study. JAMA Cardiol. **3**, 693–702 (2018)
23. Zhuang, P., Liu, X., Li, Y., Wan, X., Wu, Y., et al.: Effect of diet quality and genetic predisposition on hemoglobin A1c and Type 2 diabetes risk: gene-diet interaction analysis of 357,419 individuals. Diabetes Care **44**, 2470–2479 (2021)
24. Raschka, S.: MLxtend: providing machine learning and data science utilities and extensions to Python's scientific computing stack. J. Open Source Softw. **3**, 638 (2018)
25. Tan, P.N., Steinbach, M., Kumar, V.: Introduction to data mining. https://www-users.cse.umn.edu/~kumar001/dmbook/dmsol_11_07_2021.pdf. Accessed 8 July 2022
26. Heye, A.K., et al.: Blood pressure and sodium: association with MRI markers in cerebral small vessel disease. J. Cereb. Blood Flow Metab. **36**, 264–274 (2016)
27. Strazzullo, P., D'Elia, L., Kandala, N.-B., Cappuccio, F.P.: Salt intake, stroke, and cardiovascular disease: meta-analysis of prospective studies. BMJ **339**, b4567 (2009)
28. Santisteban, M.M., Iadecola, C.: Hypertension, dietary salt and cognitive impairment. J. Cereb. Blood Flow Metab. **38**, 2112–2128 (2018)
29. Moser, V.A., Pike, C.J.: Obesity and sex interact in the regulation of Alzheimer's disease. Neurosci. Biobehav. Rev. **67**, 102–118 (2016)

30. Rocca, W.A., Mielke, M.M., Vemuri, P., Miller, V.M.: Sex and gender differences in the causes of dementia: a narrative review. Maturitas **79**, 196–201 (2014)
31. Podcasy, J.L., Epperson, C.N.: Considering sex and gender in Alzheimer disease and other dementias. Dialog. Clin. Neurosci. **18**, 437–446 (2016)
32. Udeh-Momoh, C., Watermeyer, T.: Female Brain Health and Endocrine Research (FEMBER) consortium: female specific risk factors for the development of Alzheimer's disease neuropathology and cognitive impairment: call for a precision medicine approach. Ageing Res. Rev. **71**, 101459 (2021)
33. Bocancea, D.I., van Loenhoud, A.C., Groot, C., Barkhof, F., van der Flier, W.M., Ossenkoppele, R.: Measuring resilience and resistance in aging and Azheimer disease using residual methods: a systematic review and meta-analysis. Neurology **10** (2021)
34. Marmarelis, M.G., Ver Steeg, G., Jahanshad, N., Galstyan, A.: Bounding the effects of continuous treatments for hidden confounders (2022). http://arxiv.org/abs/2204.11206

Learning Interpretable Regularized Ordinal Models from 3D Mesh Data for Neurodegenerative Disease Staging

Yuji Zhao[1], Max A. Laansma[2], Eva M. van Heese[2], Conor Owens-Walton[3], Laura M. Parkes[4], Ines Debove[5], Christian Rummel[5], Roland Wiest[5], Fernando Cendes[6], Rachel Guimaraes[6], Clarissa Lin Yasuda[6], Jiun-Jie Wang[7], Tim J. Anderson[8], John C. Dalrymple-Alford[15], Tracy R. Melzer[8], Toni L. Pitcher[8], Reinhold Schmidt[9], Petra Schwingenschuh[9], Gäetan Garraux[10], Mario Rango[11], Letizia Squarcina[11], Sarah Al-Bachari[4], Hedley C. A. Emsley[4], Johannes C. Klein[12], Clare E. Mackay[12], Michiel F. Dirkx[13], Rick Helmich[13], Francesca Assogna[14], Fabrizio Piras[14], Joanna K. Bright[3], Gianfranco Spalletta[14], Kathleen Poston[3], Christine Lochner[16], Corey T. McMillan[17], Daniel Weintraub[17], Jason Druzgal[18], Benjamin Newman[18], Odile A. Van Den Heuvel[2], Neda Jahanshad[3], Paul M. Thompson[3], Ysbrand D. van der Werf[2], Boris Gutman[1(✉)], and for the ENIGMA consortium

[1] Illinois Institute of Technology, Chicago, USA
bgutman1@iit.edu
[2] Amsterdam UMC, Vrije Universiteit Amsterdam, Amsterdam, Netherlands
[3] University of Southern California, Los Angeles, USA
[4] The University of Manchester, Manchester, England
[5] Inselspital, University of Bern, Bern, Switzerland
[6] UNICAMP Universidade Estadual de Campinas, Campinas, Brazil
[7] Chang Gung University, Taoyuan City, Taiwan
[8] University of Otago, Dunedin, New Zealand
[9] Medical University Graz, Graz, Austria
[10] University of Liège, Liège, Belgium
[11] Fondazione IRCCS, University of Milan, Milan, Italy
[12] University of Oxford, Oxford, England
[13] Radboud University, Nijmegen, Netherlands
[14] IRCCS Santa Lucia Foundation, Rome, Italy
[15] New Zealand Brain Research Institute, University of Canterbury, Christchurch, New Zealand
[16] Stellenbosch University, Stellenbosch, South Africa
[17] University of Pennsylvania, Philadelphia, USA
[18] University of Virginia, Charlottesville, USA

Abstract. We extend the sparse, spatially piecewise-contiguous linear classification framework for mesh-based data to ordinal logistic regression. The algorithm is intended for use with subcortical shape and cortical thickness data where progressive clinical staging is available, as is generally the case in neurodegenerative diseases. We apply the tool to Parkinson's and Alzheimer's disease staging. The resulting biomarkers predict

A. Abdulkadir et al. (Eds.): MLCN 2022, LNCS 13596, pp. 115–124, 2022.
https://doi.org/10.1007/978-3-031-17899-3_12

Hoehn-Yahr and cognitive impairment stages at competitive accuracy; the models remain parsimonious and outperform one-against-all models in terms of the Akaike and Bayesian information criteria.

Keywords: Ordinal regression · Imaging biomarker · Neurodegenerative Disease

1 Introduction

Imaging biomarkers of neurodegenerative disease (NDD) continue to generate substantial interest in the clinical neurology community. MRI in particular is useful for pre-symptomatic detection of characteristic atrophy patterns [22]. Generally, practitioners prefer quantitative markers that are accurate and neuroscientifically interpretable [10]. These two competing demands - especially the interpretability requirement - have somewhat limited the clinical adoption of modern machine learning-based imaging markers, with substantial efforts in making deep learning models of NDD interpretable [6,19]. Mesh-based data extracted from MR imaging, including anatomical shape markers, have been shown to be especially useful [22] - but interpreting e.g. deep convolutional networks of mesh-based features is even more difficult. To ensure a balance between interpretability and accuracy, substantial efforts have focused on regularized linear models [3,7]. These can be extended to mesh-based data, e.g. when using sparse spatially contiguous (TV-L1) priors in NDD classification [23].

Further, as NDDs are progressive diseases, clinicians expect markers to accurately identify the stage of the disease, as exemplified by Braak staging [1] or the Alzheimer's Disease (AD) Amyloid Cascade Hypothesis [9]. Binary classifiers cannot satisfy this need. In response, a family of generative time-dependent models - termed "Disease Progression Models" (DPM) - have been developed for use with neuroimaging data [5,11,14,18]. In this typically unsupervised framework, the time (or disease stage) variable is itself inferred by optimally "stitching" together short-term individual measurements to reconstruct biomarker trajectories. While DPMs offer a means to bring several modalities together into an interpretable dynamic marker, they often require longitudinal data and may suffer from stability issues. This makes DPMs useful in guiding our understanding of disease dynamics, but limits their use in stage inference of individuals.

To address these challenges here, we return to the supervised discriminative framework. We assume the existence of clinical symptom severity measures that are related by some strictly increasing (or strictly decreasing) function to a quantitative imaging marker. In general, clinical staging cannot be expected to be related by a constant linear rate to an imaging biomarker. This makes ordinal regression ideally suited for developing a progressive discriminative biomarker, as used in its naïve form for AD stage prediction [4]. Towards this, we extend the TV-L1 regularization and optimization techniques used in [23] to ordinal logistic regression (ordit) on mesh-based features. We apply the mesh-based TV-L1 ordit to a large multi-cohort study of Parkinson's Disease (PD) study to predict

Hoehn-Yahr stage using subcortical shape. We also apply the tool to AD stage prediction in the ADNI dataset, using subcortical shape and cortical thickness. Our results outperform one-against-all classification in terms of Bayesian and Akaike information criteria, with competitive multi-class accuracy, while providing a readily interpretable model visualization.

2 Methods

2.1 Discrete Ordinal Labels in Neurodegenerative Disease

Discrete, ordered measures of clinical symptom severity are ubiquitous in NDD [15]. For example, in the context of Alzheimer's Disease, clinicians use the following labels to broadly group afflicted and at-risk patients: Alzheimer's Disease (AD), (early and late) Mild Cognitive Impairment (MCI), and normal/healthy cognition (HC). Similarly the 5 Hoehn and Yahr (H-Y) stages in Parkinson Disease [8] are commonly used to assess the severity of a patient's Parkinsonism. Although these categories are ordered, their numerical value cannot be treated as a continuous variable. An imaging biomarker will generally have a non-linear monotonic relationship with ordinal severity measures.

2.2 Ordinal Logistic Regression

Unlike binary or one-vs-all multi-class models, an ordit model simultaneously optimizes a single latent space embedding *and* monotonic between-class boundaries within this space to maximally separate classes with a known ordering.

Suppose label y can assume K distinct ordered values. Ordinal classification modifies the binary case by introducing $K-1$ boundaries $\theta_1 < \theta_2 < \theta_3 < ... < \theta_{K-1}$ to construct the following loss function [20]:

$$Loss_o(z;y) = \sum_{l=1}^{K-1} f(s(l;y)(\theta_l - z)),\ s(l;y) = \begin{cases} -1 & \text{if } l < y \\ +1 & \text{if } l \geq y \end{cases} \tag{1}$$

where z is the linear predictor $z = \beta' x$ and f is the margin penalty function. In the case of ordinal logistic regression, f is the logistic loss.

2.3 TV-L1 Regularized Regression

Predictive modeling with sparse Total Variation (TV-L1) regularization has been widely used in functional and diffusion MR brain imaging [3,17], and more recently for regression and spectral analysis of mesh-based data [2]. Here we use TV-L1-regularized Ordinal Logistic Regression (Ordit-TVL1), whose loss is

$$Loss(\beta, D, \omega_{L1}, \omega_{TV}) = Loss_o(\beta, D) + \omega_{L1}|\beta| + \omega_{TV}\sum_i ||\nabla\beta_i|| \tag{2}$$

where $Loss_o$ is the ordinal logistic loss, ω_{L1}, ω_{TV} are regularization weights, and $D = \{x_i, y_i\}$ are the data and class labels. x_i here denotes a feature vector including vertices from cortical or subcortical surface models. Finding the optimal linear coefficients requires us to solve a non-smooth convex optimization problem. Two of the most popular solvers are Alternating Direction Method of Multipliers (ADMM) and methods based on proximal operators of the non-smooth terms [3]. Here, we use an implementation based on Nesterov's smoothing. In this solver, the TV term, for which a closed-form proximal operator does not exist, is approximated with a differentiable function, while allowing exact optimization of the L1 term [2].

3 Experiments

3.1 Data

We used T1-weighted MR images from the ADNI (Alzheimer's) and ENIGMA-PD (Parkinson's) datasets. ADNI: our cohort consisted of 195 AD, 398 MCI, and 225 HC subjects. ENIGMA-PD: this is a collaborative Parkinson's Disease imaging consortium, comprised of over 20 cohorts. Figure 1 describes ENIGMA-PD clinical and demographic characteristics. We treated H-Y stages 3, 4 and 5 as one stage due to the limited number of samples in the latest three stages. This was also deemed neuroscientifically sensible, as the jump from stage 2 to 3 is considered to be most clinically significant. Our vertex-wise features consisted of cortical thickness extracted with FreeSurfer in ADNI, and subcortical medial shape thickness - used in both datasets - extracted from seven bilateral regions with a previously described approach [21].

	N	Age	% Female
PD	**2525***	**63.4 ± 9.7**	**34.8%**
Control	**1326**	**59.5 ± 12.4**	**44.7%**
HY1	436	59.5 ± 9.9	39.5
HY2	1047	64.5 ± 9.1	32.5
HY345	339	65.6 ± 9.9	41.0

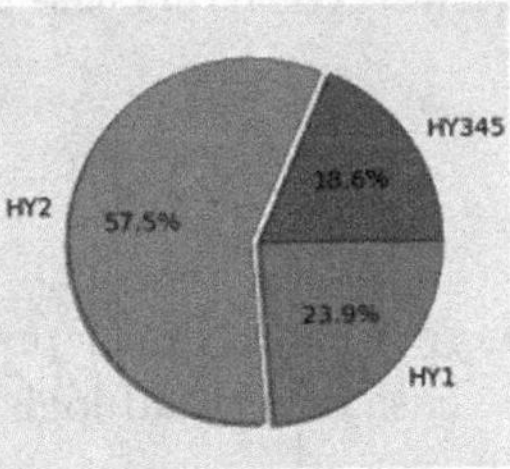

HY2 57.5%, HY345 18.6%, HY1 23.9%

Fig. 1. ENIGMA-PD dataset clinical diagnosis. * - Some PD subjects did not have HY stage information.

3.2 Experimental Pipeline

In all experiments, we used 4-fold nested cross-validation, with inner folds used to optimize TV and L1 hyper-parameters (ω_{L1}, ω_{TV}) for best balanced accuracy. Since ROC area-under-the-curve (AUC) is not appropriate for evaluating multi-class models, we used balanced f1 scores and overall balanced accuracy.

As a baseline reference, we trained relevant binary logit models with TV-L1 regularization (Logit-TVL1): AD vs. NC for the Alzheimer's study and PD vs. NC for the Parkinson's study. As the primary reference model in subcortical experiments, we also trained a conventional one-vs-all multi-class logit classifier, again with TV-L1 terms ("One-vs-All-TVL1"). Beyond the measures above, we also compared the multi-class models' parsimony and intepretability more directly, using Akaike and Bayesian information criterion (AIC, BIC). We note that for biological reasons explained below, we excluded control subjects from multi-class Parkinson's experiments, while training additional relevant binary models.

4 Results

In ADNI subcortical experiments, Logit-TVL1 model achieved ROC-AUC score of 0.91 for HC vs. AD. As expected (Fig. 2(A,B)), linear weights of the two binary models and the ordinal model show broadly consistent patterns. However, the ordinal model shows a noticeable deviation in the canonical progression pattern, with more subtle patterns in the bilateral thalamus and hippocampus. Specifically, the ordit model focuses more on the mediodorsal and pulvinar nuclei. In cortical thickness models (Fig. 3), the binary and ordinal maps also generally resemble each other. However, the ordinal map shows that the classifier is more focused on the medial temporal regions (esp. parahippocampal gyrus), and superior frontal areas, and less on lateral temporal regions. This is broadly consistent with known atrophy and amyloid/tau accumulation patterns in early stages of the disease [12], probably indicating that the model is most focused on the MCI group. We note that positive (red) weights in a multivariate linear classifier map do not necessarily imply disease-associated increase in gray matter. Just as likely, warm colors represent regions where atrophy is correlated with but reduced compared to the atrophy in the more severely affected regions (blue). As a sanity check, we trained both cortical and subcortical ordinal models *without regularization*. In both cases, the model simply failed to converge with below chance classification accuracy at its last iteration.

The ADNI multi-task classification performance of One-vs-All-TVL1 and Ordit-TVL1 models are displayed in Fig. 5. The Ordit-TVL1 model achieved 0.51 overall out-of-fold accuracy, compared with the One-vs-All-TVL1 accuracy of 0.45. The ordinal model results in a more balanced confusion matrix (Fig. 4) - given a chance accuracy of 0.33 - and balanced F1 scores across all diagnostic categories (Fig. 5). It implies that the Ordit-TVL1 model is advantageous to deal with unbalanced data in the multi-class setting. One the other hand, the One-vs-All-TVL1 maps (Fig. 2(C)) do not lend themselves easily to interpretation, as the set of binary problems does not necessarily reflect disease progression. This is further reflected in AIC and BIC scores: The one-vs-all model achieved AIC of 43733 and BIC of 54666. For comparison, the ordit-TVL1 model achieved AIC of 25294 and BIC of 37944 (lower A/BIC imply greater parsimony).

In ENIGMA-PD experiments, the baseline binary model achieved AUC of 0.65, with similar results for between stage binary classification (Fig. 6(B).

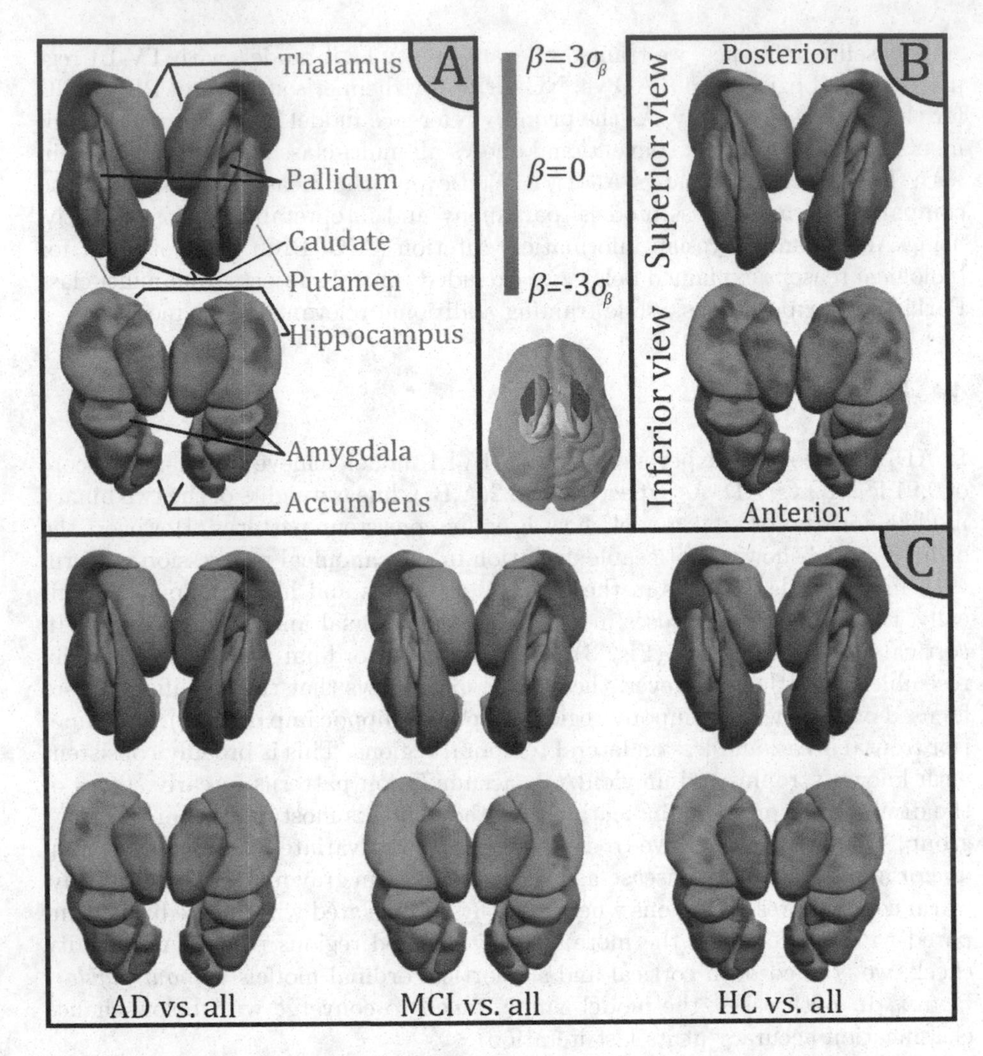

Fig. 2. Weight maps of subcortical shape models of Alzheimer's Disease. A. Binary Logit-TVL1 model separating AD and HC subjects; **B.** Ordit-TVL1 model; **C.** Components of the One-vs-All-TVL1 model. The colorbar range is defined by the standard deviation of linear loadings over all vertices, σ_β. Note the more focused degenerative pattern in the ordinal model's thalamus and hippo maps in contrast to (A). Compare the single progression map of (B) to the convoluted interpretation of (C).

Comparing the ordit model (Fig. 6(A) to the multi-class model, we again see improved A/BIC scores: Ordit-TVL1's 36552 (BIC), 24369 (AIC) versus One-vs-All-TVL1's 63681 (BIC), 42455 (AIC). The balanced f1 score and accuracy were (f1 = 0.44, acc = 0.45) for Ordit-TVL1 and (f1 = 0.45, acc = 0.46) for Logit-TVL1 (Fig.7). The linear models maps in Fig. 6 show patterns of general thinning in the striatal and limbic areas, and a "flattening" of the caudate

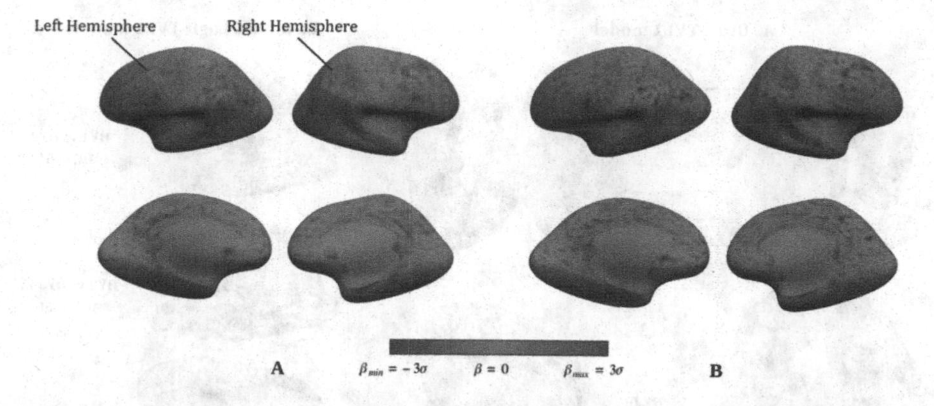

Fig. 3. Weight maps of cortical thickness models of Alzheimer's Disease. A. Logit-TVL1 model separating AD vs. HC; **B.** Ordit-TVL1 model. The grey patch in the bottom row (medial view) corresponds to vertices with no GM values (corpus callosum, ventricular boundaries, etc.).

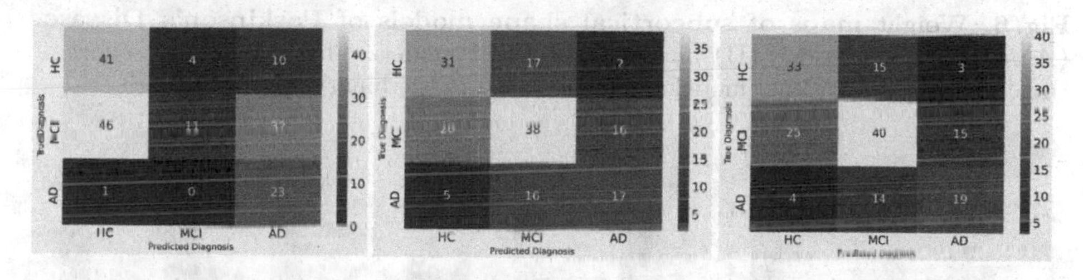

Fig. 4. Alzheimer's Multi-task classification confusion matrix. Left: Subcortical One-vs-All-TVL1 model; **Middle:** Subcortical Ordit-TVL1 model (middle); **Right:** Cortical Ordit-TVL1 model.

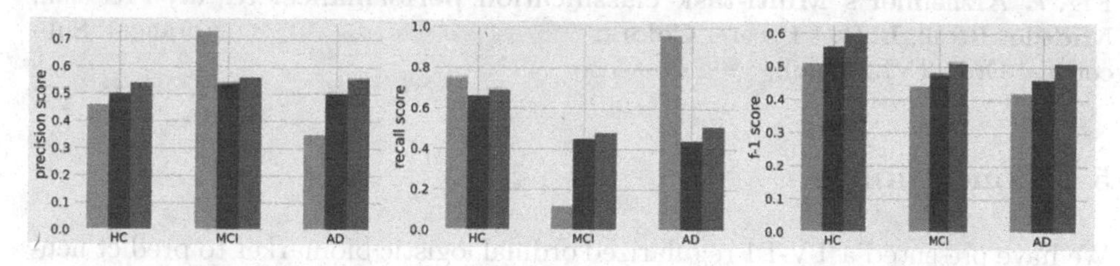

Fig. 5. Alzheimer's Multi-task classification performance. Right: Precision; **Middle:** Recall; **Left:** F1 score. **Colors:** Subcortical One-vs-All TVL1(orange), Subcortical Ordit-TVL1(blue), Cortical Ordit-TVL1 (red).

nucleus. This is consistent with previous studies of shape in PD pointing towards subcortical degeneration in these regions [16]. The thalamus forms a notable exception, being thicker in early-stage PD, while reverting to thinning in later HY stages. This is in line with subcortical volume work by ENIGMA-PD [13], and is the reason control subjects were excluded from multi-class PD models here.

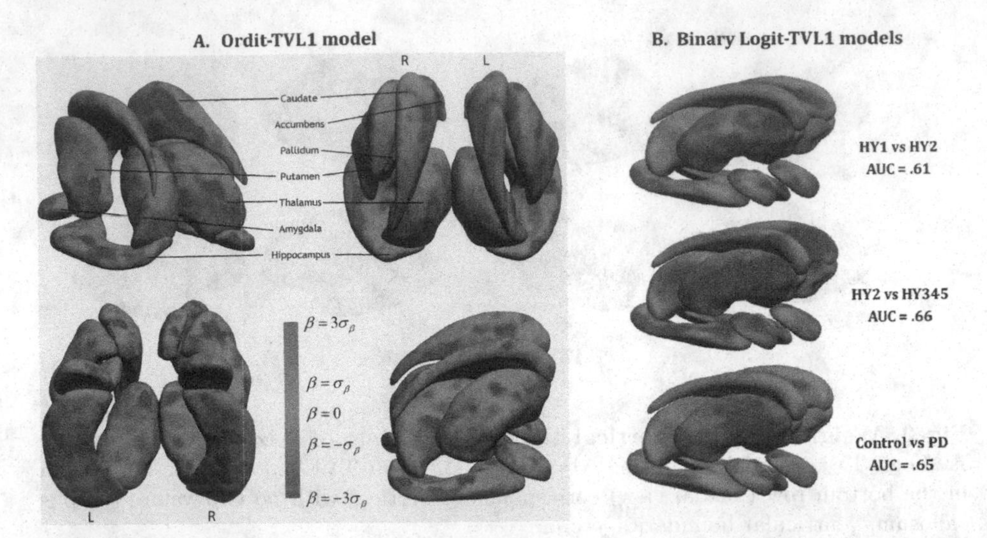

Fig. 6. Weight maps of subcortical shape models of Parkinson's Disease. **(A)** Ordit-TVL1 model, **(B)** binary Logit-TVL1 models. The "pancaking" effect of the caudate in the ordinal model is also observed in the binary PD-HC map.

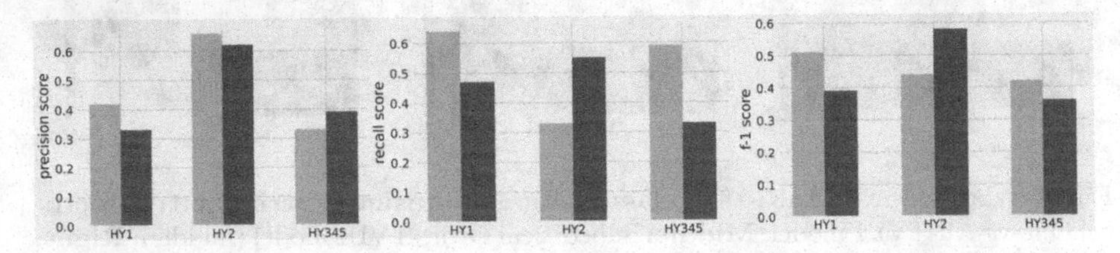

Fig. 7. Alzheimer's Multi-task classification performance. Right: Precision; **Middle:** Recall; **Left:** F1 score. **Colors:** Subcortical One-vs-All TVL1(orange), Subcortical Ordit-TVL1(blue).

5 Conclusion

We have presented a TV-L1 regularized ordinal logistic biomarker to predict neurodegenerative disease stage. The biomarker retains balanced accuracy competitive with previously reported results in AD, while remaining sparse and neuroscientifically interpretable. The improved balance of parsimony and performance is confirmed by a favorable information criteria comparison with standard one-against-all approaches. Future extensions may include incorporating regularized ordit models in multiple-task learning [23] and semi-supervised cross-disorder stage identification [11].

Acknowledgments. Work by BG and YZ was supported by the Alzheimer's Association grant 2018-AARG-592081, Advanced Disconnectome Markers of Alzheimer's Disease. ENIGMA-PD (YW, PT, EH, ML) is supported by NINDS award 1RO1NS107513-01A.

References

1. Braak, H., Del Tredici, K., Rüb, U., de Vos, R.A.I., Jansen Steur, E.N.H., Braak, E.: Staging of brain pathology related to sporadic parkinson's disease. Neurobiology of Aging **24**(2), 197–211 (2003)
2. de Pierrefeu, A.: Structured sparse principal components analysis with the tv-elastic net penalty. IEEE Trans. Med. Imaging **37**(2), 396–407 (2018)
3. Dohmatob, E.D., Gramfort,A., Thirion, B., Varoquaux, G.: Benchmarking solvers for tv-l1 least-squares and logistic regression in brain imaging. In: 2014 International Workshop on Pattern Recognition in Neuroimaging, pp. 1–4 (2014)
4. Doyle, O.M., et al.: Predicting progression of alzheimer's disease using ordinal regression. PLoS ONE **9**(8), e105542 (2014)
5. Garbarino, S., Lorenzi, M.: Modeling and inference of spatio-temporal protein dynamics across brain networks. In: Chung, A.C.S., Gee, J.C., Yushkevich, P.A., Bao, S. (eds.) IPMI 2019. LNCS, vol. 11492, pp. 57–69. Springer, Cham (2019). https://doi.org/10.1007/978-3-030-20351-1_5
6. Guo, X., Tinaz, S., Dvornek, N.C.: Characterization of early stage parkinson's disease from resting-state fmri data using a long short-term memory network. Front. Neuroimaging **1** (2022)
7. Gutman, B.A.: Empowering imaging biomarkers of Alzheimer's disease. Neurobio. Aging **36**, S69–S80 (2014)
8. Hoehn, M.M., Yahr, M.D., et al.: Parkinsonism: onset, progression, and mortality. Neurology **50**(2), 318–318 (1998)
9. Jack Jr., C.R., et al.: Hypothetical model of dynamic biomarkers of the Alzheimer's pathological cascade. Lancet Neurol. **9**(1), 119–128 (2010)
10. Jin, D., et al.: Generalizable, reproducible, and neuroscientifically interpretable imaging biomarkers for Alzheimer's disease. Adv. Sci. (Weinh) **7**(14), 2198–3844 (2020)
11. Kurmukov, A., Zhao, Y., Mussabaeva, A., Gutman, B.: Constraining disease progression models using subject specific connectivity priors. In: Schirmer, M.D., Venkataraman, A., Rekik, I., Kim, M., Chung, A.W. (eds.) CNI 2019. LNCS, vol. 11848, pp. 106–116. Springer, Cham (2019). https://doi.org/10.1007/978-3-030-32391-2_11
12. La Joie, R.: Prospective longitudinal atrophy in alzheimer&x2019;s disease correlates with the intensity and topography of baseline tau-pet. Sci. Trans. Med. **12**(524), eaau5732 (2020)
13. Laansma, M.A., et al.: International multicenter analysis of brain structure across clinical stages of Parkinson's disease. Mov. Disord. **36**(11), 2583–2594 (2021)
14. Marinescu, R.V., et al.: Dive: a spatiotemporal progression model of brain pathology in neurodegenerative disorders. Neuroimage **192**, 166–177 (2019)
15. McCullagh, P.: Regression models for ordinal data. J. Roy. Stat. Soc.: Ser. B (Methodol.) **42**(2), 109–127 (1980)
16. Nemmi, F., Sabatini, U., Rascol, O., Péran, P.: Parkinson's disease and local atrophy in subcortical nuclei: insight from shape analysis. Neurobiol. Aging **36**(1), 424–433 (2015)

17. Nir, T.M., et al.: Alzheimer's disease classification with novel microstructural metrics from diffusion-weighted MRI. In: Fuster, A., Ghosh, A., Kaden, E., Rathi, Y., Reisert, M. (eds.) Computational Diffusion MRI. MV, pp. 41–54. Springer, Cham (2016). https://doi.org/10.1007/978-3-319-28588-7_4
18. Oxtoby, N.P.: Data-driven sequence of changes to anatomical brain connectivity in sporadic Alzheimer's disease. Front. Neuro. **8**, 580 (2017)
19. Shangran, Q., et al: Development and validation of an interpretable deep learning framework for alzheimer's disease classification. Brain **143**(6), 1920–1933 (2020)
20. Rennie, J.D.M., Srebro, N.: Loss functions for preference levels: Regression with discrete ordered labels. In: Proceedings of the IJCAI Multidisciplinary Workshop on Advances in Preference Handling, vol. 1. Citeseer (2005)
21. Roshchupkin, G.V., Gutman, B.A., et al.: Heritability of the shape of subcortical brain structures in the general population. Nat. Commun. **7**, 13738 (2016)
22. Young, P.N.E., et al.: Imaging biomarkers in neurodegeneration: current and future practices. Alzheimers Res. Ther. **12**(1), 49 (2020)
23. Zhao, Y., Kurmukov, A., Gutman, B.A.: Spatially adaptive morphometric knowledge transfer across neurodegenerative diseases. In: 2021 IEEE 18th International Symposium on Biomedical Imaging (ISBI), pp. 845–849 (2021)

Augmenting Magnetic Resonance Imaging with Tabular Features for Enhanced and Interpretable Medial Temporal Lobe Atrophy Prediction

Dongsoo Lee[1], Chong Hyun Suh[2(✉)], Jinyoung Kim[1], Wooseok Jung[1], Changhyun Park[1], Kyu-Hwan Jung[1], Seo Taek Kong[1], Woo Hyun Shim[2], Hwon Heo[3], and Sang Joon Kim[2]

[1] VUNO Inc., Seoul, Republic of Korea
dslee@vuno.co

[2] Department of Radiology and Research Institute of Radiology, Asan Medical Center, University of Ulsan College of Medicine, Seoul, Republic of Korea
chonghyunsuh@amc.seoul.kr

[3] Department of Convergence Medicine, Asan Medical Center, University of Ulsan College of Medicine, Seoul, Republic of Korea

Abstract. Medial temporal lobe atrophy (MTA) score is a key feature for Alzheimer's disease (AD) diagnosis. Diagnosis of MTA from images acquired using magnetic resonance imaging (MRI) technology suffers from high inter- and intra-observer discrepancies. The recently-developed Vision Transformer (ViT) can be trained on MRI images to classify MTA scores, but is a "black-box" model whose internal working is unknown. Further, a fully-trained classifier is also susceptible to inconsistent predictions by nature of its labels used for training. Augmenting imaging data with tabular features could potentially rectify this issue, but ViTs are designed to process imaging data as its name suggests. This work aims to develop an accurate and explainable MTA classifier. We introduce a multi-modality training scheme to simultaneously handle tabular and image data. Our proposed method processes multi-modality data consisting of T1-weighted brain MRI and tabular data encompassing brain region volumes, cortical thickness, and radiomics features. Our method outperforms various baselines considered, and its attention map on input images and feature importance scores on tabular data explains its reasoning.

Keywords: Vision transformer · Explainability · Multi-modality

1 Introduction

Medial temporal lobe atrophy (MTA) score is highly related to Alzheimer's Disease (AD) [6,15]. Even expert raters often disagree and inter/intra observer variation is high when annotating MTA in magnetic resonance imaging (MRI) data [19]. Therefore, deep learning models rating MTA score can assist human raters to improve consistency if provided with accurate and explainable prediction.

A. Abdulkadir et al. (Eds.): MLCN 2022, LNCS 13596, pp. 125–134, 2022.
https://doi.org/10.1007/978-3-031-17899-3_13

A combination of convolutional neural network (CNN) and recurrent neural network (RNN) is applicable to MTA classification [11]. Moreover, their method provides a explainable visualization map using SmoothGrad [21]. However, this work has some limitations. Even though the MTA score is a discrete label, their model was trained using a loss function that is used for continuous labels and had insufficient classification performance. A promising extension is to augment radiography with tabular features to train a CNN as in [18]. The joint use of imaging and tabular data is sought in medical applications [17,22] because radiography may not contain enough information for diagnosis and such critical clinical information is mostly provided in a tabular form. Moreover, MRI's intensity or texture is highly affected by vendors and imaging protocols. Such information can be also represented in a tabular form and then properly considered during training. However, joint use of imaging and tabular data has several challenges. Dimension difference between imaging and tabular data makes it difficult to combine them [9]. Fusing multiple data can also cause overfitting as a consequence of redundant features [25].

Multi-modality data has been used to enhance CNNs in medical applications. The Alzheimer's disease (AD) classification task uses combined MRI and positron emission tomography (PET) data to improve the CNN-based model's prediction performance [4]. In cancer histology, survival CNN integrates histology images and genomic biomarkers to predict time-to-event outcomes, and such a model enhances prediction accuracy for the overall survival of patients [13]. With the recent development of Vision Transformer (ViT) [5] where long-term dependencies can be learnt with less inductive bias [10,23], we seek to apply this propitious architecture to MTA classification.

The focus of this work is in classifying MTA from both imaging and tabular data and improving its explainability. To this end, we propose an explainable multi-modality training scheme to learn feature representation from both image and tabular data in an end-to-end fashion. A ViT model provides a relevance map based on layer-wise relevance propagation (LRP) [3] and TabNet [1] can effectively learn representations from extracted image features and tabular data. Our proposed method outperforms existing methods and provides clear explainability for both image and tabular data.

2 Methods

We present a novel framework to simultaneously learn feature representation from image and tabular data. The proposed method does not only capture contextual information from the image but also effectively encodes auxiliary data in a tabular form. Further, it improves model interpretability by highlighting relevant regions and important tabular features in a predictive task. We employed two independent transformer models: ViT and TabNet [1].

A vision transformer block consists of an embedding layer, transformer block, and fully connected layer. Our transformer block consists of 4 transformer layers that include the layer normalization, attention block, and multi-layer perceptron (MLP) block and the transformer block has 512 dimensions and 4 heads.

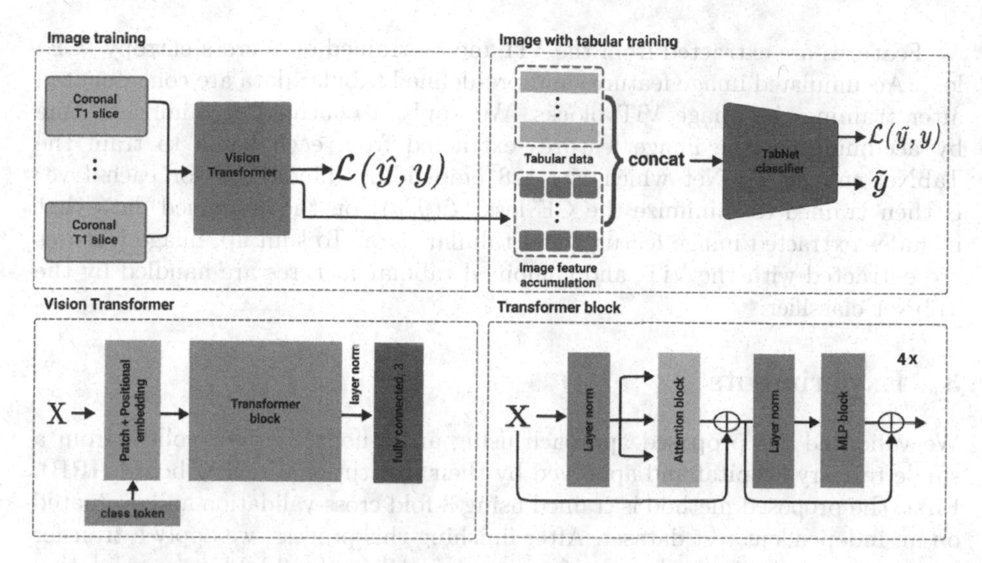

Fig. 1. Overview of the proposed multi-modality training scheme. One ViT model extracts image features from coronal T1 slices and image features are accumulated until the end of the ViT's one epoch training. Then, image features are merged with pre defined tabular data, and a TabNet classifier processes this merged data. Our ViT model consists of a positional embedding layer, a transformer block that contains 4 transformer layers, and a final fully connected layer with 3 dimensions.

We use a 4-layer ViT which is shallow compared to its typical usage because the dataset used in this work is smaller than standard public datasets used for natural imaging and such model can prevent overfitting. The ViT model can produce a global visualization feature map for input MRI data. This visualization map [3] is produced by integrating LRP-based relevance propagation that is calculated by deep Taylor decomposition [14] and gradients information. This integrated relevance map can be reached the input level and control non-linearity functions that include skip-connection and self-attention layers.

A TabNet classifier is designed to handle tabular data. The TabNet encoder is composed of a feature transformer, attentive transformer, and feature masking block that uses sparsemax function [12]. The transformer modules are the main components in the TabNet. The feature masking block can provide local interpretability which shows how the features are integrated and their influence, and global interpretability that visualizes the importance of each feature to the trained model from this sparsity. The TabNet handles raw tabular data without any pre-processing steps. Our TabNet uses jointly extracted image features and tabular data. This classifier quantifies the importance of each image or tabular feature. Tabular data includes demographics, quantitative brain region volume, cortical thickness, and radiomics information. Our proposed approach named a multi-modality training scheme is shown in Fig. 1.

Features are extracted from the ViT model trained on a cross entropy (CE) loss. Accumulated image features and pre-defined tabular data are concatenated after training the image ViT blocks. We apply a epoch-wise training scheme by accumulating the image features extracted from each batch to train the TabNet model. TabNet which has 128 embedding dimensions for each layer is then trained to minimize the CE loss ($\mathcal{L}(\tilde{y}, y)$) on the combined data that includes extracted image features and tabular data. To sum up, image features are extracted with the ViT, and combined tabular features are handled by the TabNet classifier.

3 Experiments

We validated our proposed approach using an in-house dataset collect from a single tertiary hospital and approved by their institutional review board (IRB). First, the proposed method is trained using 3-fold cross-validation and evaluated on an independent test dataset. After finishing this process, we apply a transfer learning method for the best performance model in the 3-fold cross-validation to boost our model's classification performance by using an independent second training set. Six-hundred forty three subjects were used for the cross-validation. One hundred and fifty subjects were considered for the test set. One thousand and two-hundred subjects were utilized for the transfer learning. Subjects consecutively visited a tertiary hospital for cognitive impairment. Training and validation sets in the cross-validation were matched with sex, age, and MTA scores distribution using the propensity score matching (PSM) method [2]. The MTA score was manually assessed by two experienced neuroradiologists. MTA scores are typically classified into five levels between 0 and 4. We clustered the MTA score into 3 groups (0,1) (2) (3,4) by clinically significance which reduces ambiguity arising from the subjective nature of MTA scores. Abnormal subjects are classified as MTA ≥ 2 for subjects under 75 years of age and ≥ 3 for those above [20]. Table 1 summarizes each dataset's distributions.

Table 1. Dataset statistics for each data set.

	Cross-validation	Transfer	Test
Number of Subjects	643	1200	150
Sex (male/female)	250/393	462/738	56/94
Age (Mean ± Std)	69.21 ± 10.21	72.34 ± 9.01	68.49 ± 10.99
MTA group (0/1/2)	211/152/280	295/449/456	50/50/50

3.1 Data Preprocessing

3D T1-weighted MRI scans were processed using Free Surfer [7]: scans were resampled into $256 \times 256 \times 256$ voxels with an spatial resolution of $1 \times 1 \times 1$ mm and registered to a fixed MRI template using rigid transformation. Brain skulls were then removed. Fifty coronal slices focusing on MTA regions were resized in the dimension 224×224 and these were used as ViT inputs. The slices have been commonly used as the region of interest (ROI) for MTA visual rating [20].

For tabular data, we conducted brain region segmentation using an in-house deep learning-based segmentation model and extracted volume, cortical thickness, and radiomics for each brain region using pyradiomics [24]. Extraction of radiomics features requires both the image and corresponding mask of region of interest. Our segmented maps are exploited to extract radiomics features and we focus on the shape features that are independent from intensity values, thus, these features can be extracted from raw image dataset. The shape feature can additionally capture morphological information in the brain region. Dimension of tabular features in each image was reduced, considering its significance based on the chi-square test. We selected the 30 features that represent the significant differences between the MTA groups. Then, we pre-trained these features using the TabNet and chose final tabular features using the TabNet's feature importance. We left features that had non-zero importance and included sex, age, slice index features. The slice index means the location of the MRI slice in the coronal view. Interestingly, final tabular data (17 features) are closely related with medial temporal lobe regions [16]: [**Volume**: inferior lateral ventricle, hippocampus, X3rd ventricle, choroid plexus, lateral ventricle, cerebrospinal fluid], [**Cortical thickness**: middle temporal, precuneus], [**Radiomics**: (Left hippocampus - surface volume ratio, mesh volume, voxel volume) (Right hippocampus - surface volume ratio, mesh volume), X3rd ventricle flatness], sex, age, slice index.

3.2 Evaluation

For inference, we used the middle 30 slices out of 50 slices used per training subject and conducted hard-voting classification using each slice per subject. We utilized the mode of the predicted results for each of the 30 slices to classify subjects for hard-voting. We compared the proposed method with various models: (i) ViT using only 3D T1-weighted MRI images; (ii) TabNet using only tabular information; (iii) ViT with one fully-connected (FC) layer, where the extracted latent features are concatenated with tabular data and processed through a FC layer. (iv) ViT with TabNet but without accumulating extracted image features (v) EfficientNetV2-small with TabNet.

We used the Adam and sharpness-aware minimization (SAM) optimizer [8] to train the models. Our initial learning rate was 0.001 and SAM's base optimizer was stochastic gradient descent (SGD) with momentum of 0.9. We used cosine annealing warm restarts learning scheduler to adjust learning rate ($T_0 : 5, T_{mult} : 1, \eta_{\min} : 1e-6$) and trained for 35 epochs. We calculated the performance on the test set using the best evaluation model on the validation set and ensembled the result of each fold. We used precision, recall, and F1-score that were calculated by the macro-average method because there was no class imbalance in our test set. All models and training schemes were implemented by PyTorch (version 1.7.1), and experiments were conducted on the NVIDIA GeForce GTX 1080 Ti GPU.

3.3 Results

Table 2. Classification performance for the external test set. Only TabNet uses tabular data alone, only ViT uses 3D T1-weighted MRI alone. ViT + 1FC, ViT + TabNet, EfficientNetV2-small + TabNet, and ours exploit both tabular and MR image data.

Model	Precision	Recall	F1-score
Only TabNet (SAM) - baseline	0.759	0.767	0.754
Only TabNet (Adam) - baseline	0.752	0.760	0.752
Only ViT (SAM) - baseline	0.757	0.760	0.755
Only ViT (Adam) - baseline	0.341	0.427	0.325
ViT + 1FC (SAM)	0.775	0.780	0.768
ViT + 1FC (Adam)	0.788	0.780	0.761
ViT + TabNet (SAM)	0.781	0.787	0.778
EfficientNetV2-small + TabNet (SAM)	0.768	0.773	0.769
Ours (SAM)	**0.801**	**0.787**	**0.790**
Ours (SAM) - Transfer learning	**0.817**	**0.800**	**0.804**

Table 2 summarizes the classification performance of various models and our proposed model. Only TabNet and only ViT are uni-modality models. The former model used particularly tabular data and the latter model used image data alone. We found that there was no significant difference in predictive performance between only TabNet and only ViT. Also, TabNet's performance was unaffected by the choice of optimizer between SAM and Adam, but Adam failed to successfully train the only ViT model. When extracted image features and tabular features were combined, the predictive performance was improved by approximately 0.02 in all metrics (see ViT + 1FC model, ViT + TabNet). This result indicates that combining image and tabular features can boost classification performance. Ours has superior performance than only TabNet (by 0.042 precision, 0.02 recall, and 0.036 F1-score) and only ViT (by 0.044 precision, 0.027 recall, and 0.035 F1-score). We also conducted a homogeneity test to compare the classification performance between ours and unimodality baseline models.

Ours performance was better than the baselines and the differences were statistically significant ($p = 0.049$ between ours and only ViT, $p = 0.008$ between ours and only TabNet). This result proved that multi-modality could enhance the model's achievement. Moreover, our proposed method outperformed ViT + 1FC (by 0.026 precision, 0.007 recall, and 0.022 F1-score) and ViT + TabNet (by 0.020 precision and 0.012 F1-score). This result explains that the proposed multi-modality cascaded training scheme was more effective than simply concatenating image and tabular features with the FC layer or TabNet. Cross-validation results for ours were 0.830 for precision, 0.821 for recall, and 0.825 for F1-score. As can be seen in this case, the performance of ours was not highly deteriorated when applied to the external test set. Similarly, EfficientNetV2-small, the CNN-based state-of-the-art model failed to exceed the our proposed model's performance. Rather, CNN-based model yielded inferior performance than the ViT model. Then, the transfer learning model outperformed ours (by 0.016 precision, 0.013 recall, 0.014 F1-score) and this demonstrates that adding training data through transfer learning can further improve model's performance. We also found that our method better-classified cases with the MTA group of 1. Identifying MTA group 1 is considered to be more challenging than classifying MTA groups 0 or 2 because group 1 has ambiguity due to their middle grade location.

Table 3. Top-5 feature importance for each fold when applied to test data set. Abbreviations: Volume (V), Radiomics (R), Inferior lateral (IL), Left hippocampus (LH), Right hippocampus (RH), Surface volume ratio (SVR).

Feature importance	Fold 1	Fold 2	Fold 3
Rank 1	IL ventricles (V)	IL ventricles (V)	IL ventricles (V)
Rank 2	Hippocampus (V)	LH voxel volume (R)	Hippocampus (V)
Rank 3	RH SVR (R)	Sex	RH mesh volume (R)
Rank 4	Sex	RH SVR (R)	RH SVR (R)
Rank 5	LH SVR (R)	LH SVR (R)	Slice index

Our proposed model and only ViT (SAM) can provide attention maps for model interpretability while others fail to visualize such information. Note that training of only ViT (Adam) failed, and thus its visualization map is not meaningful. However, our method used cascaded approach by training the ViT and TabNet models to avoid such an effect. We demonstrate that our proposed method outperformed other baseline methods and obtained clear interpretable visualization maps that precisely highlight the MTA relevant regions (Fig. 2). Furthermore, our TabNet model can identify influential tabular features and therefore provide interpretable results as shown in Table 3.

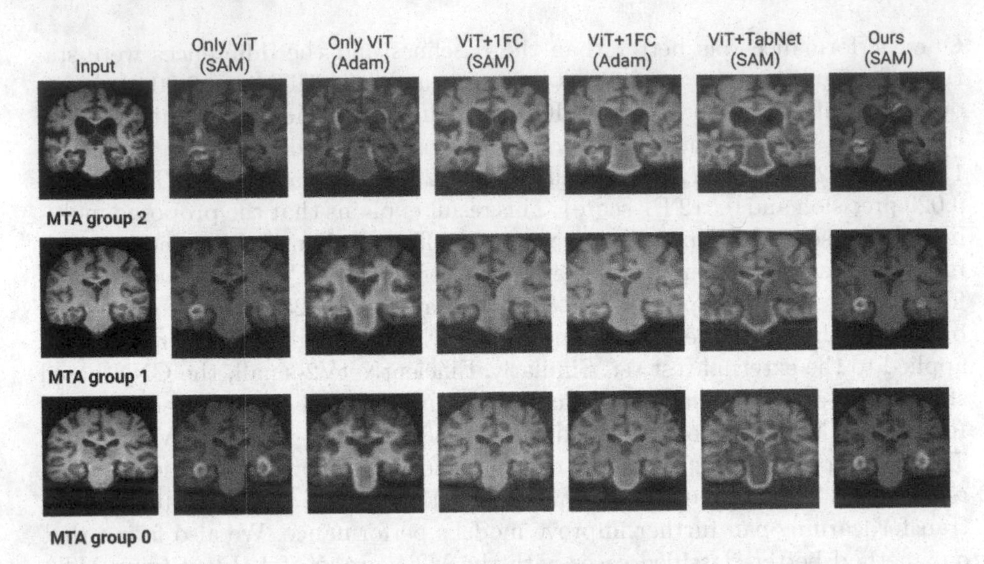

Fig. 2. The input image of each row is an MRI slice representing the MTA group 2,1,0 in order. Each column shows the visualization map for each model for the corresponding slice. As shown in the figure, our method produces more accurate visualization maps. The first-row map (MTA group 2) indicates the region of the MTA and ventricle. The second (MTA group 1) and third row (MTA group 0) point to the MTA region.

4 Conclusion

In this paper, we proposed a novel framework that can exploit multi-modality data consisting of image and tabular features when training transformer-based models. We trained transformer-based models using multi-modality data while keeping a clear interpretable visualization map and comparable prediction performance. Our experiments showed that our method outperforms previous approaches and other baselines that combine image and tabular data in MTA group prediction. Moreover, we demonstrated that transfer learning can further improve the model's performance.

Future works include development of independent MTA score classification models for each left and right using the original MTA score definition by cropping the specific medial temporal lobe region. Furthermore, we will extend our proposed model to solve a multi-tasking problem for MTA score prediction and Alzheimer's disease diagnosis.

References

1. Arık, S.O., Pfister, T.: Tabnet: attentive interpretable tabular learning. In: AAAI, vol. 35, pp. 6679–6687 (2021)
2. Caliendo, M., Kopeinig, S.: Some practical guidance for the implementation of propensity score matching. J. Econ.Surv. **22**(1), 31–72 (2008)

3. Chefer, H., Gur, S., Wolf, L.: Transformer interpretability beyond attention visualization. In: Proceedings of the IEEE/CVF Conference on Computer Vision and Pattern Recognition, pp. 782–791 (2021)
4. Cheng, D., Liu, M.: Cnns based multi-modality classification for ad diagnosis. In: 2017 10th International Congress on Image and Signal Processing, Biomedical Engineering and Informatics (CISP-BMEI), pp. 1–5. IEEE (2017)
5. Dosovitskiy, A., et al.: An image is worth 16×16 words: Transformers for image recognition at scale. arXiv preprint arXiv:2010.11929 (2020)
6. Duara, R., et al.: Medial temporal lobe atrophy on mri scans and the diagnosis of Alzheimer disease. Neurology **71**(24), 1986–1992 (2008)
7. Fischl, B.: Freesurfer. Neuroimage **62**(2), 774–781 (2012)
8. Foret, P., Kleiner, A., Mobahi, H., Neyshabur, B.: Sharpness-aware minimization for efficiently improving generalization. arXiv preprint arXiv:2010.01412 (2020)
9. Lahat, D., Adali, T., Jutten, C.: Multimodal data fusion: an overview of methods, challenges, and prospects. Proc. IEEE **103**(9), 1449–1477 (2015)
10. Liu, Z., et al.: Swin transformer: hierarchical vision transformer using shifted windows. In: Proceedings of the IEEE/CVF International Conference on Computer Vision, pp. 10012–10022 (2021)
11. Mårtensson, G., et al.: Avra: automatic visual ratings of atrophy from mri images using recurrent convolutional neural networks. NeuroImage Clin. **23**, 101872 (2019)
12. Martins, A., Astudillo, R.: From softmax to sparsemax: a sparse model of attention and multi-label classification. In: International Conference on Machine Learning, pp. 1614–1623. PMLR (2016)
13. Mobadersany, P., et al.: Predicting cancer outcomes from histology and genomics using convolutional networks. Proc. Natl. Acad. Sci. **115**(13), E2970–E2979 (2018)
14. Montavon, G., Lapuschkin, S., Binder, A., Samek, W., Müller, K.R.: Explaining nonlinear classification decisions with deep taylor decomposition. Pattern Recogn. **65**, 211–222 (2017)
15. Park, H.Y., Park, C.R., Suh, C.H., Shim, W.H., Kim, S.J.: Diagnostic performance of the medial temporal lobe atrophy scale in patients with Alzheimer's disease: a systematic review and meta-analysis. Eur. Radiol. **31**(12), 9060–9072 (2021)
16. Park, Y.W., et al.: Radiomics features of hippocampal regions in magnetic resonance imaging can differentiate medial temporal lobe epilepsy patients from healthy controls. Sci. Rep. **10**(1), 1–8 (2020)
17. Pölsterl, S., Sarasua, I., Gutiérrez-Becker, B., Wachinger, C.: A wide and deep neural network for survival analysis from anatomical shape and tabular clinical data. In: Cellier, P., Driessens, K. (eds.) ECML PKDD 2019. CCIS, vol. 1167, pp. 453–464. Springer, Cham (2020). https://doi.org/10.1007/978-3-030-43823-4_37
18. Pölsterl, S., Wolf, T.N., Wachinger, C.: Combining 3D image and tabular data via the dynamic affine feature map transform. In: de Bruijne, M., et al. (eds.) MICCAI 2021. LNCS, vol. 12905, pp. 688–698. Springer, Cham (2021). https://doi.org/10.1007/978-3-030-87240-3_66
19. Scheltens, P., Launer, L.J., Barkhof, F., Weinstein, H.C., van Gool, W.A.: Visual assessment of medial temporal lobe atrophy on magnetic resonance imaging: interobserver reliability. J. Neurol. **242**(9), 557–560 (1995)
20. Scheltens, P., et al.: Atrophy of medial temporal lobes on MRI in" probable" Alzheimer's disease and normal ageing: diagnostic value and neuropsychological correlates. J. Neurol. Neurosurg. Psych. **55**(10), 967–972 (1992)
21. Smilkov, D., Thorat, N., Kim, B., Viégas, F., Wattenberg, M.: Smoothgrad: removing noise by adding noise. arXiv preprint arXiv:1706.03825 (2017)

22. Spasov, S., et al.: A parameter-efficient deep learning approach to predict conversion from mild cognitive impairment to Alzheimer's Disease. Neuroimage **189**, 276–287 (2019)
23. Touvron, H., Cord, M., Douze, M., Massa, F., Sablayrolles, A., Jégou, H.: Training data-efficient image transformers & distillation through attention. In: International Conference on Machine Learning, pp. 10347–10357. PMLR (2021)
24. Van Griethuysen, J.J., et al.: Computational radiomics system to decode the radiographic phenotype. Can. Res. **77**(21), e104–e107 (2017)
25. Wang, W., Tran, D., Feiszli, M.: What makes training multi-modal classification networks hard? In: Proceedings of the IEEE/CVF Conference on Computer Vision and Pattern Recognition, pp. 12695–12705 (2020)

Automatic Lesion Analysis for Increased Efficiency in Outcome Prediction of Traumatic Brain Injury

Margherita Rosnati[1,2,3](✉), Eyal Soreq[2,3], Miguel Monteiro[1], Lucia Li[2,3], Neil S. N. Graham[2,3], Karl Zimmerman[2,3], Carlotta Rossi[4], Greta Carrara[4], Guido Bertolini[4], David J. Sharp[2,3], and Ben Glocker[1]

[1] BioMedIA Group, Department of Computing, Imperial College London, London, UK
margherita.rosnati12@imperial.ac.uk
[2] UK Dementia Research Institute Care Research and Technology Centre, Imperial College London, London, UK
[3] Department of Brain Sciences, Faculty of Medicine, Imperial College London, London, UK
[4] Laboratory of Clinical Epidemiology, Istituto di Ricerche Farmacologiche Mario Negri IRCCS, Bergamo, Italy

Abstract. The accurate prognosis for traumatic brain injury (TBI) patients is difficult yet essential to inform therapy, patient management, and long-term after-care. Patient characteristics such as age, motor and pupil responsiveness, hypoxia and hypotension, and radiological findings on computed tomography (CT), have been identified as important variables for TBI outcome prediction. CT is the acute imaging modality of choice in clinical practice because of its acquisition speed and widespread availability. However, this modality is mainly used for qualitative and semi-quantitative assessment, such as the Marshall scoring system, which is prone to subjectivity and human errors. This work explores the predictive power of imaging biomarkers extracted from routinely-acquired hospital admission CT scans using a state-of-the-art, deep learning TBI lesion segmentation method. We use lesion volumes and corresponding lesion statistics as inputs for an extended TBI outcome prediction model. We compare the predictive power of our proposed features to the Marshall score, independently and when paired with classic TBI biomarkers. We find that automatically extracted quantitative CT features perform similarly or better than the Marshall score in predicting unfavourable TBI outcomes. Leveraging automatic atlas alignment, we also identify frontal extra-axial lesions as important indicators of poor outcome. Our work may contribute to a better understanding of TBI, and provides new insights into how automated neuroimaging analysis can be used to improve prognostication after TBI.

A. Abdulkadir et al. (Eds.): MLCN 2022, LNCS 13596, pp. 135–146, 2022.
https://doi.org/10.1007/978-3-031-17899-3_14

1 Introduction

Traumatic brain injury (TBI) is a leading cause of death in Europe [27]. In the UK alone, 160,000 patients with TBI are admitted to hospitals annually, with an estimated yearly cost of £15 billion [21]. The accurate prediction of TBI outcome is still an unresolved challenge [9,13] whose resolution could improve the therapy and after-care of patients. Due to its acquisition speed and wide availability, computed tomography (CT) is the imaging modality of choice in clinical practice [14] and a key component in TBI outcome prediction [4]. The Marshall score [17] is one of the most widely used metrics to evaluate TBI injury severity. However, it does not leverage the rich information content of CT imaging [2] and requires a radiologist to assess the CT scan manually, which is time-consuming. Years of acquisition of CT of TBI patients generated rich datasets, opening the possibility of automatic extraction of CT biomarkers. These could enable a deeper and broader use of imaging data, augmenting the skills of radiologists and reducing the workload, allowing them to see more patients. Machine learning for medical imaging is a growing research field with advances in medical imaging segmentation [25] and classification [6]. It can be used for fast and autonomous outcome prediction of TBI using imaging data.

This work explores the predictive power of novel TBI biomarkers computationally extracted from hospital admission CT scans. TBI lesion volumes are automatically extracted from the scans; then, lesion statistics are derived to inform the prediction of TBI outcome. We compare the discriminate power of our proposed features to the Marshall score features independently and when paired with clinical TBI biomarkers.

In particular, we make the following contributions:

- **Novel machine-learning driven imaging biomarkers.** We extract interpretable measurements for TBI lesion quantification;
- **Human-level performance on unfavourable outcome prediction.** We reach comparable, if not superior, performance to manually extracted CT biomarkers when predicting unfavourable outcome, both in isolation and when paired with clinical TBI biomarkers;
- **Imaging biomarker relevance.** We show that features relating to extra-axial haemorrhage in the frontal lobe are important for the prediction of outcome, confirming previous clinical findings in a data-driven manner.

2 Related Work

The prediction of TBI of outcome has primarily been tackled using clinical features. Jiang et al. [12] and Majdan et al. [15] used clinical features such as age and motor score to predict the patient outcome with a regression model. Pasipanodya et al. [22] focused on predictions for different patient subgroups. Huie et al. [10] provide an extensive review of clinical biomarkers for the prediction of unfavourable outcome in TBI. More recently, Bruschetta et al. [3] and Matsuo et al. [18] investigated using the same predictors with machine learning models, such as support vector machines and neural networks. Researchers also used more complex features to predict TBI outcome, such as electroencephalograms [20] and magnetic resonance imaging (MRI) [8].

A stream of research within TBI focuses on features extracted from CT. Recent work adopted neural networks for TBI lesion segmentation and midline shift quantification [11,19].

Similarly to our work, Plassard et al. [24] and Chaganti et al. [5] used multi-atlas labelling to extract radiomics from different brain regions and predict a variety of TBI end-points, yet excluding unfavourable outcome. Pease et al. [23] trained a convolutional neural network using CT scans and clinical TBI biomarkers to predict the TBI outcome and achieved comparable performance to IMPACT. Unlike the biomarkers we designed, the authors extracted deep learning features, which are not interpretable by humans. In parallel to our work, Yao et al. [28] trained a neural network to segment epidural haematomas and used the resulting segmentation volumes to predict patient mortality. Our work differs because we focused on the more challenging and clinically relevant problem of predicting TBI unfavourable outcome. TBI outcomes are measured by the patient state scale Glasgow Outcome Scale Extended (GOS-E), where a score of 4 or below defines an unfavourable outcome.

3 Methods

Study Design. We analysed data from the observational studies The Collaborative REsearch on ACute Traumatic Brain Injury in intensiVe Care Medicine in Europe (CREACTIVE, NCT02004080), and BIOmarkers of AXonal injury after TBI (BIO-AX-TBI, NCT03534154 [7]), the latter being a follow-up study to CREACTIVE. From these observational studies, sub-studies collecting CT scans recruited 1986 patients admitted to intensive care with a diagnosis of TBI between 2013 and 2019 from 21 European hospitals. CREACTIVE recruited patients admitted to a hospital due to suspected TBI, whereas for BIO-AX-TBI, the criterion was moderate-severe TBI as defined in [16]. The studies did not define a protocol for acquisition or scanner type. Hence, the CT scans collected were a heterogeneous dataset akin to clinical scenarios.

Data Cleaning. We discarded patients for whom trauma time, hospital admission and discharge, admission CT scan, and the patient outcome measure GOS-E were not recorded. In addition, we discarded patients who had surgery before the scan, whose scan could not be automatically aligned to an atlas, and for whom one of the clinical TBI biomarkers - age, Glasgow Coma Scale motor score and pupil responsiveness, hypoxia and hypotension - or Marshall score was missing. A flowchart of the patient selection can be found in the appendix Fig. 5. Of the remaining 646 patients, 389 (60.2%) had an unfavourable outcome, of which 225 (34.8%) died. The median age is 56.6, and 74.9% of the patients are male. Out of the 21 centres, we selected three centres with 59 patients as an independent replication holdout test set, and we used the remaining 18 centres with 587 patients as a training set.

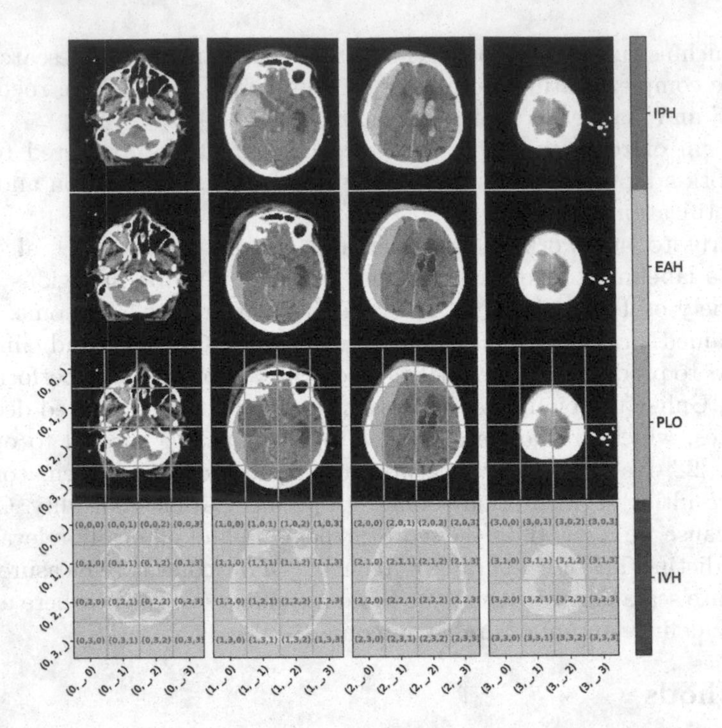

Fig. 1. Example of feature extraction. A CT scan (1st row) is fed through a segmentation neural network, outputting lesion volumes (2nd row + legend). We either aggregated the lesion volumes globally or calculated the volumes over each cuboidal region to provide localised measures (3rd and 4th rows).

Feature Extraction. All CT scans were standardised via automatic atlas registration to obtain rigid alignment to MNI space. We then used the deep-learning lesion segmentation method, BLAST-CT, described in [19]. The method produces a voxel-wise classification of intraparenchymal haemorrhages (IPH), extra-axial haemorrhages (EAH), perilesional oedemas (PLO) and intraventricular haemorrhages (IVH), and reports respective Dice similarity coefficients of 65.2%, 55.3%, 44.8% and 47.3% for lesions above 1 mL.

From the 3D segmentations, we extracted two types of statistical features, *global* and *local* lesion biomarkers. We calculated all connected components for each lesion type and discarded any connected region of 0.02 mL or less to remove noise. We determined this threshold by visual inspection of the scans. From the separate lesions[1], we extracted the following *global* lesion biomarkers: the total number of lesions, the median lesion volume, the 25th and 75th lesion volume percentiles, and the mean lesion volume. Next, we partitioned the registered 3D scans four ways for each axis, creating $4^3 = 64$ equidimensional cuboids. We chose four partitions to balance feature granularity and human interpretability. We extracted *local* lesion biomarkers for each lesion type by calculating the

[1] Each lesion is a connected component.

Table 1. Models performance when predicting unfavourable outcome. Marshall score is described in [17], *global* features refer to disjoint lesion statistics, and *local* features refer to lesion volume per cuboid (see **Feature extraction**). The clinical biomarkers are age, motor and pupil responsiveness, hypoxia and hypotension.

Features	Cross-validation AUROC	Hold-out set AUROC	Hold-out set Precision	Hold-out set Recall
Marshall score	76.7 +/- 7.7	73.2	69.6	61.5
global lesion features	72.4 +/- 5.8	80.9	70.0	**80.8**
local lesion features	77.2 +/- 5.5	83.3	65.6	**80.8**
***local* + *global* lesion features**	**77.2 +/- 6.5**	**84.0**	**74.1**	76.9
Features	**Cross-validation AUROC**	**Hold-out set AUROC**	**Hold-out set Precision**	**Hold-out set Recall**
Marshall score + clinical biomarkers	81.6 +/- 3.9	84.7	69.2	69.2
global + clinical biomarkers	82.1 +/- 4.3	87.5	80.8	**80.8**
***local* + clinical biomarkers**	**83.0 +/- 4.4**	**87.7**	**84.0**	**80.8**
local + *global* + clinical biomarkers	81.1 +/- 4.8	87.2	77.8	**80.8**

total lesion volume in each cuboidal region. Figure 1 shows an example of brain partitioning and the corresponding indexing.

Modelling and Performance Evaluation. We used a Random Forest Classifier with 300 estimators to predict a patient's favourable or unfavourable outcome. An unfavourable outcome, defined as a GOS-E score of 4 or below, is a typical target in TBI outcome prediction. We compared eight predictive models based on different sets of features. The first four models use imaging features alone: 1) the Marshall score; 2) *global* lesion biomarkers; 3) *local* lesion biomarkers; 4) *global* and *local* lesion biomarkers. The second set of models uses the imaging features above, together with the clinical TBI biomarkers (age, motor and pupil responsiveness, hypoxia and hypotension). Note that the clinical TBI biomarkers and Marshall score are the same as those used in the state-of-the-art IMPACT-CT model [26].

We evaluated model performance using the area under the receiver-operator curve (AUROC), precision, recall, and true positive rate at a 10% false positive rate. In addition to evaluating performance on the holdout test set, we also measured cross-validation performance on the training set. We calculated statistical significance through a permutation test on the holdout set's metrics and the statistical relevance of each feature using the average Gini importance. In addition, a cross-validation per clinical centre can be found in appendix Table 2.

4 Results

***Local* and *Global* Lesion Biomarkers Performed Similarly or Better than Marshall Score.** Using *local* and *global* lesion biomarkers achieved a cross-validation AUROC of 76.7 ± 7.7% compared to 77.0 ± 6.6% when using the

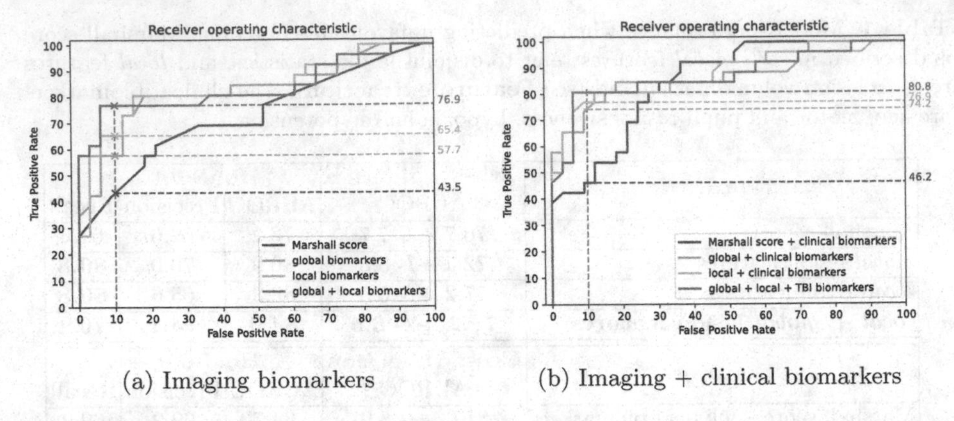

(a) Imaging biomarkers (b) Imaging + clinical biomarkers

Fig. 2. ROC curves of models predicting unfavourable outcome. The dashed lines indicate the true positive rate for a fixed false positive rate of 10%. *Global* and *local* biomarkers always produce a better or equivalent performance than the Marshall score.

Marshall score (Table 1 top). On the holdout set, the improvement of AUROC was 10.8%, from 73.2% using the Marshall score to 84.0% using *local* and *global* lesion biomarkers. Similarly, the precision improved by 4.5% and the recall by 15.4%. For a false positive rate of 10%, the volumetric features yielded a true positive rate of 73.1% compared to 43.5% for the Marshall score (Fig. 2a). Testing the statistical significance of these improvements, we found that improvement in AUROC in the holdout set was statistically significant (one-way p-value $<$ 0.05), whereas all other metrics on the holdout set were statistically comparable.

***Local* Lesion Biomarkers and Clinical Features Performed Similarly or Better than Features Used in IMPACT-CT.** When adding the clinical TBI features to the *local* lesion biomarkers, the AUROC was 83.0 $\pm$ 4.4% in cross-validation and 87.7% on the holdout set, compared to 81.6 $\pm$ 3.9% and 84.7%, respectively, for clinical TBI and Marshall score biomarkers (Table 1 bottom). Similarly, the holdout set's precision, recall and true-positive rate improved by 3%, 14.8% and 34.6%, respectively (Fig. 2b). When tested, the improvement in true positive rate was significant (one-way p-value < 0.05), whereas the remaining holdout set metrics for the two experiments were statistically comparable.

Extra-Axial Haemorrhage was the Most Important Feature. When considering feature importance when using *global* lesion biomarkers (Fig. 3b), EAH was the statistical feature with the highest importance score, and IVH was the feature with the lowest. The lesion count and maximum lesion size were the most important factors. As per the *global* lesion biomarkers, EAH was the statistical feature with the highest importance scores when using *local* lesion biomarkers to predict unfavourable outcome (Fig. 4). In addition IVH in the second-bottom transverse plane, second-bottom coronal plane $((1, 2, \_))$, EAH in the second-front coronal plane $((1, 1, \_)$ and $(2, 1, \_))$ were important.

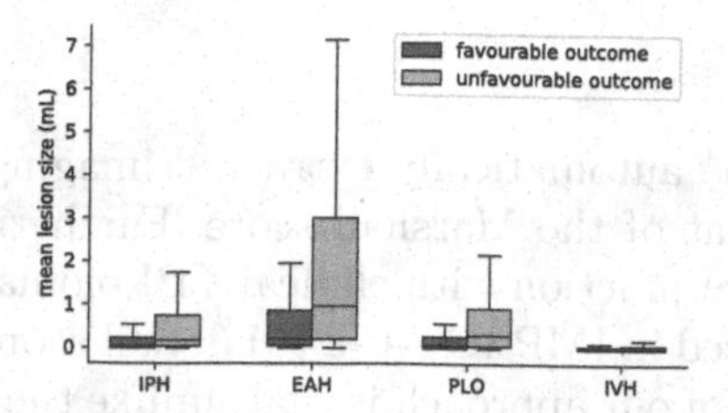

(a) Mean lesion size stratified per lesion type and outcome. Unfavourable outcome can be qualified by larger average IPH, EAH and PLO.

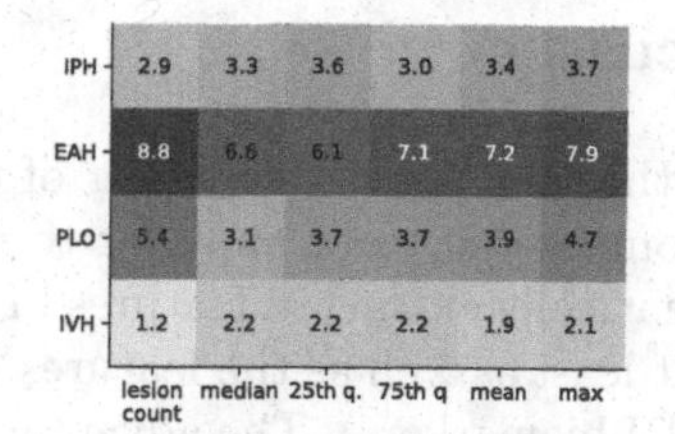

(b) Feature importance of *global* lesion biomarkers when predicting unfavourable outcome. The numbers and colour intensity refer to the Gini importance (rescaled by 10^{-2}).

Fig. 3. *Global* feature statistics and their importance for outcome prediction.

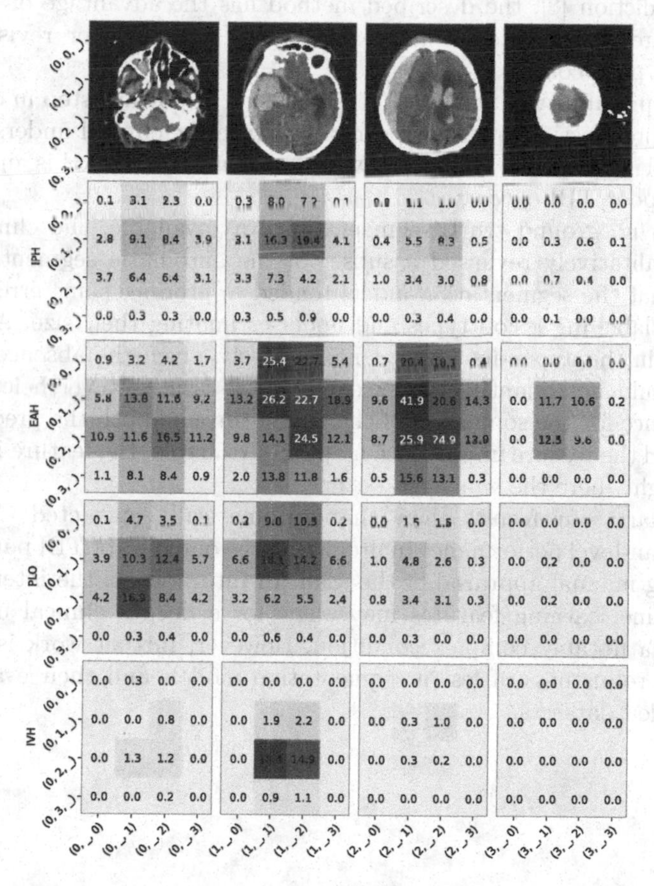

Fig. 4. Feature importance of local lesion biomarkers when predicting unfavourable outcome. Each row represents the feature importance of different lesion types, and each column represents a transversal slice, similarly to Fig. 1. For each transverse slice, each row represents a coronal slice, whereas each column represents a sagittal slice. The numbers and colour intensity refer to the Gini importance (rescaled by 10^{-3}).

5 Discussion

We show that the predictive power of the automatically extracted imaging features is comparable to or superior to that of the Marshall score. Furthermore, the automatically extracted features in conjunction with clinical TBI biomarkers perform at least as well as the features used in IMPACT-CT (Marshall score and clinical TBI biomarkers). The advantage of our approach is that, unlike the Marshall score, automatically extracted imaging features do not require a radiologist to manually review the scan, allowing faster patient care, and reducing workload. Our method generalises well across different scanner types and acquisition protocols as shown from the consistent results on the training set cross-validation and the independent hold-out set. Although other approaches, such as advanced fluid biomarker or magnetic resonance imaging, have also shown promise in improving outcome prediction [8], the described method has the advantage of using data which is currently collected routinely, obviating the need for revised clinical investigation protocols.

The interpretability of the lesion features is an important step in discovering data-driven prognostic biomarkers, contributing to the clinical understanding of TBI. Reinforcing previous results [1], we found that frontal EAH is an important indicator of poor TBI outcomes.

Although no ground truth segmentation was available, the clinicians (LL and DS) qualitatively reviewed a subset of the automatic segmentations. We concluded that the segmentation model tended to produce some errors, such as partially mislabelling lesion types and under-estimating their size. An example can be seen in the appendix Fig. 6. Unfortunately, given the absence of ground truth, we could not quantify the extent of these issues. Nevertheless, there is strong evidence for the soundness of the model through both the predictive performance and the feature importance maps. For example, the feature importance of IVH is high where the ventricles occur.

In summary, our results show that automatically extracted CT features achieve human-level performance in predicting the outcome of TBI patients without requiring manual appraisal of the scan. In future work, the interpretability of the machine learning features may allow for a deeper clinical understanding of TBI, a notably complex condition. However, further work is needed to improve the robustness of lesion segmentation models and their evaluation on new unlabelled datasets.

Acknowledgements. MR is supported by UK Research and Innovation [UKRI Centre for Doctoral Training in AI for Healthcare grant number EP/S023283/1]. LL is supported by NIHR, Academy of Medical Sciences. NSNG is funded by a National Institute for Health Research (NIHR) Academic Clinical Lectureship and acknowledges infrastructure support from the UK Dementia Research Institute (DRI) Care Research and Technology Centre at Imperial College London and the NIHR Imperial Biomedical Research Centre (BRC). The BIOAX-TBI data was collected thanks to the ERA-NET NEURON Cofund (MR/R004528/1), part of the European Research Projects on External Insults to the Nervous System call, within the Horizon 2020 funding framework. The CREACTIVE project has received funding from the European Union Seventh Framework Programme (FP7/2007–2013) under Grant Agreement number 602714.

Appendix

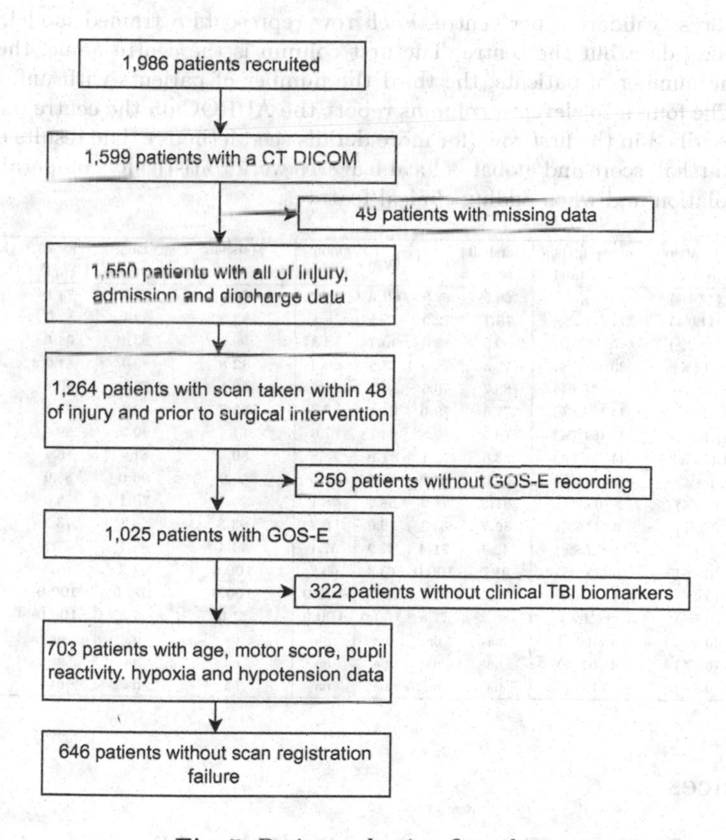

Fig. 5. Patient selection flow-chart

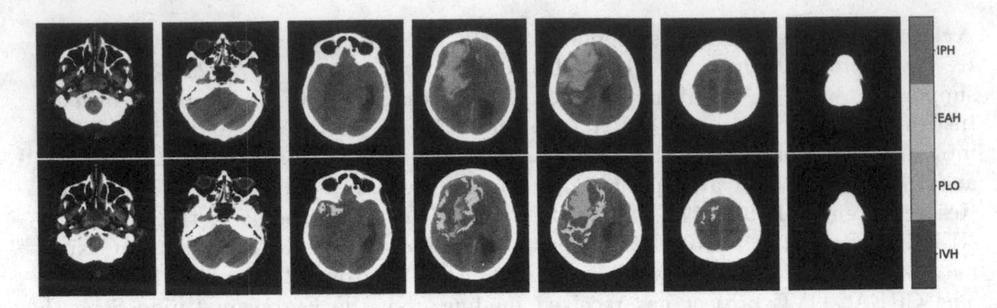

Fig. 6. Extreme example of mis-segmented CT scan. The lesion labels are intraparenchymal (IPH), extra-axial (EAH) and intraventricular (IVH) haemorrhages, and perilesional oedemas (PLO). The extra-axial label on fourth and fifth slices is erroneous; the opacity on the left side of the brain in the third slice should be labelled as oedema.

Table 2. Cross validation per centre. Each row represents a trained model, trained on all training data but the centre. The first column is the centre name, the second contains the number of patients, the third the number of patients with unfavourable outcome. The fourth to eleventh columns report the AUROC on the centre data using features described in the first row (for more details see Methods). The results obtained with the Marshall score and global + local features were statistically comparable when taken in isolation and when adding clinical features.

Centre	N. test patients (% of tot)	unfav. outcome (% of test)	Marshall score	global	local	global + local	Marshall score + clinical	global + clinical	local + clinical	global + local + clinical
SI009	80 (12.4%)	80 (56.7%)	66.5	62.8	**69.5**	68.3	**76.6**	73.8	74.1	72.8
IT079	71 (11.0%)	71 (59.2%)	**78.1**	72.0	75.8	75.1	**83.8**	83.1	82.0	78.5
IT100	79 (12.2%)	79 (79%)	90.0	85.3	**90.7**	89.3	90.9	93.4	**94.0**	93.0
IT544	30 (4.6%)	30 (51.7%)	77.3	74.0	77.5	**80.4**	**82.8**	80.4	81.6	81.7
IT064	27 (4.2%)	27 (64.3%)	83.7	70.6	**84.1**	81.1	80.1	79.5	**90.6**	85.8
IT442	14 (2.2%)	14 (66.7%)	77.6	**79.6**	62.2	67.3	**83.7**	69.4	75.5	71.4
IT062	11 (1.7%)	11 (64.7%)	**79.5**	59.8	66.7	57.6	81.8	**90.2**	86.4	84.1
IT099	11 (1.7%)	11 (64.7%)	63.6	62.1	**81.8**	78.8	**89.4**	81.8	86.4	84.8
IT651	5 (0.8%)	5 (38.5%)	55.0	**85.0**	77.5	82.5	65.0	**90.0**	85.0	87.5
IT513	8 (1.2%)	8 (66.7%)	**81.3**	71.9	68.8	78.1	70.3	**78.1**	75.0	**78.1**
IT101	3 (0.5%)	3 (37.5%)	**86.7**	53.3	80.0	66.7	**93.3**	56.7	73.3	73.3
CH001	1 (0.2%)	1 (12.5%)	42.9	**71.4**	42.9	57.1	71.4	**85.7**	57.1	57.1
IT036	6 (0.9%)	6 (85.7%)	91.7	**100.0**	83.3	66.7	**100.0**	83.3	83.3	83.3
IT034	4 (0.6%)	4 (57.1%)	95.8	91.7	**100.0**	**100.0**	**100.0**	**100.0**	**100.0**	**100.0**
IT590	3 (0.5%)	3 (50%)	83.3	77.8	**100.0**	**100.0**	55.6	77.8	**100.0**	**100.0**
IT724	4 (0.6%)	4 (100%)	na	na	na	na	na	na	na	na
IT057	4 (0.6%)	4 (100%)	na	na	na	na	na	na	na	na
IT088	2 (0.3%)	2 (100%)	na	na	na	na	na	na	na	na

References

1. Atzema, C., Mower, W.R., Hoffman, J.R., Holmes, J.F., Killian, A.J., Wolfson, A.B., National Emergency X-Radiography Utilization Study (NEXUS) II Group, et al.: Prevalence and prognosis of traumatic intraventricular hemorrhage in patients with blunt head trauma. J. Trauma Acute Care Surg. **60**, 1010-7 (2006)
2. Brown, A.W., et al.: Predictive utility of an adapted Marshall head CT classification scheme after traumatic brain injury. Brain Injury **33**, 610–617 (2019)

3. Bruschetta, R., et al.: Predicting outcome of traumatic brain injury: is machine learning the best way? Biomedicines **10**, 686 (2022)
4. Carter, E., Coles, J.P.: Imaging in the diagnosis and prognosis of traumatic brain injury. Expert Opinion Med. Diagnostics **6**, 541–554 (2012)
5. Chaganti, S., Plassard, A.J., Wilson, L., Smith, M.A., Patel, M.B., Landman, B.A.: A Bayesian framework for early risk prediction in traumatic brain injury. In: Medical Imaging 2016: Image Processing, vol. 9784. International Society for Optics and Photonics (2016)
6. Cireşan, D.C., Giusti, A., Gambardella, L.M., Schmidhuber, J.: Mitosis detection in breast cancer histology images with deep neural networks. In: Mori, K., Sakuma, I., Sato, Y., Barillot, C., Navab, N. (eds.) MICCAI 2013. LNCS, vol. 8150, pp. 411–418. Springer, Heidelberg (2013). https://doi.org/10.1007/978-3-642-40763-5_51
7. Graham, N.S.N., et al.: Multicentre longitudinal study of fluid and neuroimaging BIOmarkers of axonal injury after traumatic brain injury: the BIO-AX-TBI study protocol. BMJ Open **10**, e042093 (2020)
8. Graham, N.S., et al.: Axonal marker neurofilament light predicts long-term outcomes and progressive neurodegeneration after traumatic brain injury. Sci. Transl. Med. **13**, eabg9922 (2021)
9. Helmrich, I.R., et al.: Development of prognostic models for health-related quality of life following traumatic brain injury. Qual. Life Res. **31**(2), 451–471 (2021)
10. Huie, J.R., Almeida, C.A., Ferguson, A.R.: Neurotrauma as a big-data problem. Curr. Opinion Neurol. **31**, 702–708 (2018)
11. Jain, S., et al.: Automatic quantification of computed tomography features in acute traumatic brain injury. J. Neurotrauma **36**, 1794–1803 (2019)
12. Jiang, J.Y., Gao, G.Y., Li, W.P., Yu, M.K., Zhu, C.: Early indicators of prognosis in 846 cases of severe traumatic brain injury. J. Neurotrauma **19**, 869–874 (2002)
13. Kalanuria, A.A., Geocadin, R.G.: Early prognostication in acute brain damage: where is the evidence? Curr. Opinion Critical Care **19**, 113–122 (2013)
14. Kim, J.J., Gean, A.D.: Imaging for the diagnosis and management of traumatic brain injury. Neurotherapeutics **8**, 39–53 (2011)
15. Majdan, M., Brazinova, A., Rusnak, M., Leitgeb, J.: Outcome prediction after traumatic brain injury: comparison of the performance of routinely used severity scores and multivariable prognostic models. J. Neurosci. Rural Pract. **8**, 20–29 (2017)
16. Malec, J.F., et al.: The mayo classification system for traumatic brain injury severity. J. Neurotrauma **24**, 1417–1424 (2007)
17. Marshall, L.F., et al.: The diagnosis of head injury requires a classification based on computed axial tomography. J Neurotrauma **9**, S287–S292 (1992)
18. Matsuo, K., Aihara, H., Nakai, T., Morishita, A., Tohma, Y., Kohmura, E.: Machine learning to predict in-hospital morbidity and mortality after traumatic brain injury. J. Neurotrauma **37**, 202–210 (2020)
19. Monteiro, M., et al.: Multiclass semantic segmentation and quantification of traumatic brain injury lesions on head CT using deep learning: an algorithm development and multicentre validation study. Lancet Digital Health **2**, e314–e322 (2020)
20. Noor, N.S.E.M., Ibrahim, H.: Machine learning algorithms and quantitative electroencephalography predictors for outcome prediction in traumatic brain injury: a systematic review. IEEE Access **8**, 102075–102092 (2020)
21. Parsonage, M.: Traumatic brain injury and offending. Centre for Mental health (2016)

22. Pasipanodya, E.C., Teranishi, R., Dirlikov, B., Duong, T., Huie, H.: Characterizing profiles of TBI severity: predictors of functional outcomes and well-being. J. Head Trauma Rehabil. (2022)
23. Pease, M., et al.: Outcome prediction in patients with severe traumatic brain injury using deep learning from head CT scans. Radiology **304** (2022)
24. Plassard, A.J., Kelly, P.D., Asman, A.J., Kang, H., Patel, M.B., Landman, B.A.: Revealing latent value of clinically acquired CTs of traumatic brain injury through multi-atlas segmentation in a retrospective study of 1,003 with external cross-validation. In: Medical Imaging 2015: Image Processing, vol. 9413. International Society for Optics and Photonics (2015)
25. Ronneberger, O., Fischer, P., Brox, T.: U-Net: convolutional networks for biomedical image segmentation. In: Navab, N., Hornegger, J., Wells, W.M., Frangi, A.F. (eds.) MICCAI 2015. LNCS, vol. 9351, pp. 234–241. Springer, Cham (2015). https://doi.org/10.1007/978-3-319-24574-4_28
26. Steyerberg, E.W., et al.: Predicting outcome after traumatic brain injury: development and international validation of prognostic scores based on admission characteristics. PLoS Med. **5**, e165 (2008)
27. Tagliaferri, F., Compagnone, C., Korsic, M., Servadei, F., Kraus, J.: A systematic review of brain injury epidemiology in Europe. Acta neurochirurgica **148**, 255–268 (2006)
28. Yao, H., Williamson, C., Gryak, J., Najarian, K.: Automated hematoma segmentation and outcome prediction for patients with traumatic brain injury. Artif. Intell. Med. **107**, 101910 (2020)

Autism Spectrum Disorder Classification Based on Interpersonal Neural Synchrony: Can Classification be Improved by Dyadic Neural Biomarkers Using Unsupervised Graph Representation Learning?

Christian Gerloff[1,2,3](✉), Kerstin Konrad[1,2], Jana Kruppa[1,4], Martin Schulte-Rüther[4,5], and Vanessa Reindl[1,2,6]

[1] JARA-Brain Institute II, Molecular Neuroscience and Neuroimaging, RWTH Aachen and Research Centre Juelich, Aachen, Germany

[2] Child Neuropsychology Section, Department of Child and Adolescent Psychiatry, Psychosomatics and Psychotherapy, Medical Faculty, RWTH Aachen University, Aachen, Germany

[3] Chair II of Mathematics, Faculty of Mathematics, Computer Science and Natural Sciences, RWTH Aachen University, Aachen, Germany

christian.gerloff@rwth-aachen.de

[4] Department of Child and Adolescent Psychiatry and Psychotherapy, University Medical Center Göttingen, Göttingen, Germany

[5] Leibniz ScienceCampus Primate Cognition, Göttingen, Germany

[6] Psychology, School of Social Sciences, Nanyang Technological University, Singapore, Singapore

Abstract. Research in machine learning for autism spectrum disorder (ASD) classification bears the promise to improve clinical diagnoses. However, recent studies in clinical imaging have shown the limited generalization of biomarkers across and beyond benchmark datasets. Despite increasing model complexity and sample size in neuroimaging, the classification performance of ASD remains far away from clinical application. This raises the question of how we can overcome these barriers to develop early biomarkers for ASD. One approach might be to rethink how we operationalize the theoretical basis of this disease in machine learning models. Here we introduced unsupervised graph representations that explicitly map the neural mechanisms of a core aspect of ASD, deficits in dyadic social interaction, as assessed by dual brain recordings, termed hyperscanning, and evaluated their predictive performance. The proposed method differs from existing approaches in that it is more suitable to capture social interaction deficits on a neural level and is applicable to young children and infants. First results from functional

M. Schulte-Rüther and V. Reindl—Shared last authorship.

A. Abdulkadir et al. (Eds.): MLCN 2022, LNCS 13596, pp. 147–157, 2022.
https://doi.org/10.1007/978-3-031-17899-3_15

near-infrared spectroscopy data indicate potential predictive capacities of a task-agnostic, interpretable graph representation. This first effort to leverage interaction-related deficits on neural level to classify ASD may stimulate new approaches and methods to enhance existing models to achieve developmental ASD biomarkers in the future.

Keywords: Graph representation learning · ASD · Interbrain networks

1 Introduction

Autism spectrum disorder (ASD) is a neurodevelopmental disorder that affects approximately 1% of the population [30] and is characterized by impairments in reciprocal social interaction, communication, and repetitive stereotypic behavior [20]. Of these symptoms, deficits in social interaction are often considered as most central to the disorder [24]. Since ASD is associated with a high global burden of disease [2], early detection and intervention are of utmost importance for optimal long-term outcomes [10]. Diagnosing ASD is currently exclusively based on behavioral observation and anamnestic information. Especially at early ages, it is a challenging, time-consuming task, requiring a high level of clinical expertise and experience.

Brain imaging may provide a complementary source of information. Over the last decades, a vast body of research has documented anatomical and functional brain differences between individuals with ASD and healthy controls (e.g., [11]), however, despite increasing model complexity they failed to reveal consistent neural biomarkers. Current machine learning models predict ASD diagnosis based on functional magnetic resonance imaging data from large datasets (n>2000) with an area under the curve (AUC) of ~0.80 [26]. However, even when information leakage is avoided by adequate cross-validation (CV) procedures (e.g., [14,26]), the predictive accuracy typically decreases in medical imaging applications when validating with private hold-out sets (see [18,26,27]: from $AUC = 0.80$ to $AUC = 0.72$). Increasing sample sizes may improve this situation, however, this is often not feasible with clinical data and recent results using the largest database currently available for ASD (ABIDE) suggest an asymptotic behavior of AUC, with $AUC = 0.83$ for 10,000 ASD subjects [26]. While an $AUC = 0.83$ is promising, this is still far away from classification accuracies (AUC of up to 0.93) when applying machine learning methods to behavioral data obtained from a social interactive situation, such as ADOS [23]. One approach to overcome these roadblocks might be to rethink how we operationalize the theoretical basis of disease in machine learning models.

Although ASD is characterized by social difficulties during interaction with others [4], MRI data used for ASD classification has typically been acquired without any social interactive context. Participants lay still and alone in an MRI scanner during resting state and structural scans (e.g., [26]) and only a few studies used simple social tasks, such as passively viewing social scenarios (e.g., [8]). We suggest that a more ecologically valid neurobiological measure of social interaction could be derived from dyadic setups that simultaneously assess the brains of interacting subjects. Such "hyperscanning" settings typically use less intrusive imaging techniques, such as functional near-infrared spectroscopy (fNIRS), that are more tolerable to young children. Previous studies using a variety of tasks [1] have demonstrated that statistical dependencies of brain activity emerge across individuals, suggesting interpersonal neural synchrony (INS). Initial results in ASD suggest reduced INS during joint attention, communication, and cooperative coordination tasks ([19,25,29], but see [16] for contradictory findings). Considering that the clinical diagnosis of ASD is crucially dependent on analysing interactive, reciprocal behavior, it appears particularly promising to employ such interbrain measures for the classification of ASD.

Recently, we formalized INS using bipartite graphs, with brain regions of the two participants as two sets of nodes and inter-brain connections as edges [13]. We proposed to accommodate the non-Euclidean structures using network embeddings to predict experimental conditions from inter-brain networks. Generally, to discriminate among graphs that belong to different classes or clinical groups requires a graph representation that can either be part of an end-to-end model, such as graph neural networks (e.g., [15]), or of an encoder, e.g., Graph2Vec [17]. The former is typically trained in a supervised or semi-supervised fashion to learn class associations directly. The latter requires an intermediate step, i.e. unsupervised training to derive a task-agnostic representation which can subsequently be used for classification using a range of available algorithms. This allows to compare the predictive capacities of connectivity estimators vs. graph representations based on the same classifier. Additionally, task-related intrabrain data indicate that classifiers might benefit from lower-vector representations when sample sizes are low [6].

To summarize, we aim to contribute to the field of machine learning for clinical imaging by exploring, for the first time, unsupervised graph representation learning and its potential to contribute to the classification of ASD using a hyperscanning dataset [16]. Since such datasets are rare and typically small, this should be considered as a first step to demonstrate feasibility and to encourage 1) application of these methods to other tasks and 2) consider the collection of larger hyperscanning cohorts to further explore this approach. Specifically, we aim to contribute by:

- **Operationalization of Theoretical Basis.** We propose a new methodological approach that aims to encode a core aspect of ASD (i.e. deficits in social interaction) at the neural level by applying network embeddings on graph representations of interpersonal neural synchrony.
- **Predictive Validity.** We assessed the predictive validity to classify ASD dyads based on graph representations of interbrain networks. We examined how the employed connectivity estimator and different types of network embeddings influence classification performance.
- **Future Contribution to Early Life Biomarkers and Beyond.** The proposed method differs from existing approaches in that it i) allows to capture

social interaction deficits on a neural level and ii) is applicable to young children and infants. Moreover, we present a task-agnostic graph representation of interbrain synchrony that is applicable for ASD classification and beyond, e.g., using statistical models for inferential inquiries.

2 Methods

2.1 Problem Formulation

We formalize the problem of whether graph representations of INS can classify participants with or without ASD diagnosis.

First, given a set of graphs denoted by $X = (G_1, \ldots, G_n)$ we intend to train an encoder ϕ parameterized by Θ^E. The encoder learns unsupervised from similarities in graph space a matrix representation of all graphs $Z \in R^{|\mathbb{X}| \times \delta}$ with $Z_i \in R^{\delta}$ representing the graph $G_i \in X$. The size δ of the network embedding is either parameterized via Θ^E or given by the encoder.

Second, the predictive capacities of various ϕ are evaluated in an inductive learning regime by a classifier $\hat{C} : \mathcal{Z} \mapsto Y$ parameterized by Θ^C. The labels denoted by $Y = (y_1, \ldots, y_n)$ represent graphs with ASD ($y_i = 1$) or without ASD ($y_i = 0$). Finally, the performance metric $e(\hat{y}; y)$ of each ϕ is assessed.

2.2 Functional Connectivity

Functional connectivity estimators (FC) quantify the statistical relationship between two neural signals. They can be used for further network construction (see Definition 1) or serve as features for ASD classification (e.g., in intrabrain studies [26]). Here, we evaluate whether graph representations carry further beneficial information (e.g., topological properties) than the pure connectivity estimator. As connectivity estimators vary in the captured dynamics between signals, we calculated, based on the continuous wavelet transform, the following estimators to systematically account for different aspects of the dynamics: wavelet coherence (WCO), phase-locking value (PLV), Shannon entropy (Entropy). WCO captures mostly concurrent synchronization between two brain signals of two participants in time-frequency space. While WCO also accounts for amplitude differences, PLV considers only phase synchronization. In contrast to WCO and PLV, Entropy captures non-linear, delayed forms of synchronization.

2.3 Graph Representation Learning

While FCs describe a pairwise association between two brain regions, the estimator itself does not capture the multiple interdependencies and topological properties of the brain which has been shown to be organized and function as a network. A system formulation that accounts for these network characteristics and encodes social interaction, a core aspect of ASD, at neural level are interbrain networks [13].

Definition 1. *In accordance with [13], interbrain networks can be specified by a bipartite graph* $G = (V_1 \cup V_2, E)$ *where* V_1 *and* V_2 *denote brain regions of participant 1 and participant 2, respectively.* $E \subseteq V_1 \times V_2$ *represents the edges with the corresponding weights* W*, defined by the specific connectivity estimator.*

Based on three distinct FCs and the subsequently derived interbrain networks, we assessed the predictive capacities of state-of-the-art graph representation learning. In the following, we describe what sets each encoder apart.

NMF-Based Interbrain Network Embedding (NMF-IBNE) provides a lower-vector representation that encodes proximity of the topological properties of a bipartite graph [13]. This embedding does not assume a connected bipartite graph and can operate together with graph reduction procedures stratifying for non-interaction related connectivity. It leverages substructures based on a priori specified graph properties of interest (here, nodal density). Importantly, its basis matrix enables a direct interpretation of the contribution of each brain region due to the non-negative constraints of the NMF.

Local Degree Profile (LDP) encodes the graph structure based on the nodal degree of a node and its neighbors [7]. For this purpose, the encoder maps the nodal degree, its mean, minimum, maximum, and standard deviation of the first neighbors via a histogram or empirical distribution function.

Graph2Vec employs the neural network architecture skip-gram in graph space [17]. Conceptually, given X and a sequence of sampled subgraphs from different nodes, the algorithm minimizes log-likelihood of the rooted subgraph corresponding to a specific G. Each subgraph is derived via the Weisfeiler-Lehman relabeling process around each *node*.

GL2Vec aims to adjust Graph2Vec for an edge-centric case [9]. In the same spirit, it minimizes the likelihood of the rooted subgraph but instead of operating directly on G, it transforms each graph into a line graph. Thereby, it extracts edge-centric substructures and enhances the integration of edge properties, which sets it apart from the other approaches.

Diffusion-Wavelet-Based Node Feature Characterization (DWC) describes an algorithm that assesses the topological similarities using a diffusion wavelet and node features [28]. The eigenvalues of the Laplacian matrix describe the temporal frequencies of a signal on G. Coefficients derived from a wavelet kernel represent the energy transferred between nodes, whereas nodes with similar energy patterns have similar structural roles.

Geometric Scattering (Scattering) applies the invariant scattering transform on graphs [12]. Like DWC, the algorithm encodes topological similarities via diffusion wavelets but on the normalized Laplacian to consider nodal degree.

Feather performs random walks on the normalized adjacency matrix [22]. Assuming that the correlation between node properties is related to the role similarity, it pools the r-scaled weighted characteristic from the adjacency matrix.

2.4 Classifier

We employed two common classifiers for ASD classification, specifically, L2 regularized logistic regression and support-vector-machines (e.g., [26]).

2.5 Hyperparameter Optimization

Let $\Theta = \{\Theta^E, \Theta^C\}$ denote the model hyperparameters, we aim to select a model parametrization by splitting each training set $\mathcal{D}_{train}$ into $k_{inner} = 3$ training and distinct test sets. Hyperparameter optimization was performed via a Gaussian Process based on this cross-validation setting. The area under the curve of the receiver operating characteristic (ROC-AUC) was used as the evaluation score to estimate $\hat{\Theta} = arg\ min_{\Theta} L\left(C_{\Theta}; \mathcal{D}_{train}\right)$.

3 Experiment

3.1 Dataset

To examine the capacities of the graph representations for classifying ASD, we used the hyperscanning dataset provided by [16]. The cohort consists of 18 children and adolescents diagnosed with ASD and 18 typically developed children and adolescents, matched in age (8 and 18 years) and gender. Each child performed a cooperative and competitive computer game with the parent and an adult stranger in two task blocks. During the game, the prefrontal brain activities were measured concurrently using fNIRS. fNIRS is an optical imaging technique that measures neural activity through concentration changes in oxygenated (HbO) and deoxygenated (HbR) hemoglobin. The brain's metabolic demand for oxygen and glucose increases in active brain areas. Thereby, the concentration of HbO increases and HbR decreases. HbO and HbR were preprocessed consistently with [16]. Overall, this allows the construction of 136 ASD-related and 144 healthy control-related graphs for HbO and HbR each (see Sect. 2.2).

3.2 Evaluation Regimes

We examine the capacities to discriminate between graphs according to ASD status by performing i) CV and ii) a cross-chromophore test (CCT; see [13]). While HbO is most commonly analyzed in fNIRS studies, HbO was used in (i) and HbR served as the test set in (ii). Importantly, in (i) and (ii) unsupervised representation learning was evaluated in an inductive learning regime in which we do not expose ϕ to test data during training. In (i), we employed a nested-stratified CV to ensure a generalizable evaluation and avoid information leakage during hyperparameter optimization. To this end, X was randomly partitioned into $k_{out} = 5$ mutually exclusive subsets, where we strictly ensure that all data from one dyad is in one subset and that the proportion of ASD to healthy control dyads is preserved to ensure generalization across dyads. HbO and HbR rely on the same neurovascular coupling mechanism but have inverse characteristics. In (ii), training was performed in an isolated fashion on HbO to assess the test performance on HbR-based networks. This kind of out-of-distribution test is a unique opportunity of fNIRS to test predictive capacities across chromophores. For both evaluation regimes (i, ii), performance was quantified by ROC-AUC as an established performance metric in ASD classification [26]. For (i), mean and standard deviation across all folds were reported. In clinical settings, an evaluation of the performance variance might be particularly important (see also [5]). Thus, we employed a Bayesian correlated t-test [3] accounting for the correlation of overlapping training sets. Further, by randomly shuffling Y, we tested the robustness of the results against training on randomly labeled data.

3.3 Implementation Details and Reproducibility

Code, Parameter, and Metric Versioning. Mlflow served as a tracking and versioning environment. Parameters and performance metrics were stored in a non-relational database. The versioned code of embeddings and evaluation regimes is available in the repository: https://github.com/ChristianGerloff/IBN-ASD-classification.

Unit Tests and Technical Reproducibility. To accommodate best practices from software development, the repository includes unit tests. Dependencies were managed using Poetry. Experiments were run inside a docker container on a remote instance with 4×3.1 GHz Intel Xeon processors and 16 GiB memory.

3.4 ASD Classification Performance

Summarized, the rigorous and conservative evaluation demonstrated the challenges in ASD classification. Only results of NMF-IBNE based on Shannon entropy indicate potential capacities to predict ASD in this dataset. FC may be less robust as it suffered from particularly high variance in ROC-AUC on randomly labeled data. This might indicate that INS-based ASD classification could potentially benefit from graph representations.

Specifically, the CV results shown in Table 1 suggest that concurrent and linear forms of interaction-related synchrony (WCO, PLV) did not allow to differentiate between ASD and healthy subjects, in line with findings on population level [16]. In contrast, across both classifiers, FC and NMF-IBNE showed AUC above chance level, indicating that delayed, nonlinear forms of synchrony (Entropy) may capture ASD-related aspects of social interaction. To verify these results, a test based on randomized labels was performed (see Sect. 3.2). Models above chance level that passed this test are marked in bold in Table 1. Only

NMF-IBNE performed better than training on randomly labeled data across classifiers (HDI lies right to zero, see Fig. 1A,B), speaking for the robustness of the results. However, in this small cohort, NMF-IBNE showed only a weak tendency for increased performance of NMF-IBNE compared to FC (see Fig. 1C). Importantly, in contrast to other graph representation models, NMF-IBNE yields model intrinsic interpretability that enables to study the neural basis of ASD from an inferential perspective (see also [21]).

CCT revealed that HbO and HbR embeddings are distinct. In addition to systematic differences between HbR and HbO, both are differentially affected by physiological and motion artifacts.

Table 1. Classification performance of cross-validation and cross-chromophore test.

		WCO		PLV		Entropy	
		SVM	Ridge	SVM	Ridge	SVM	Ridge
CV	FC	0.50 ± 0.04	0.55 ± 0.09	0.53 ± 0.04	0.55 ± 0.06	0.53 ± 0.11	0.59 ± 0.07
	NMF-IBNE	0.54 ± 0.11	0.48 ± 0.07	0.52 ± 0.05	0.54 ± 0.06	**0.60** ± 0.05	**0.61** ± 0.05
	LDP	0.49 ± 0.02	0.51 ± 0.09	0.51 ± 0.07	0.43 ± 0.06	0.47 ± 0.11	0.51 ± 0.09
	Graph2Vec	0.50 ± 0.01	0.50 ± 0.01	0.52 ± 0.05	0.50 ± 0.04	0.54 ± 0.07	0.50 ± 0.01
	GL2Vec	0.53 ± 0.04	0.51 ± 0.01	0.50 ± 0.00	0.52 ± 0.03	0.49 ± 0.09	0.50 ± 0.00
	DWC	0.48 ± 0.08	0.48 ± 0.03	0.49 ± 0.02	0.47 ± 0.03	0.46 ± 0.12	0.50 ± 0.11
	Scattering	0.56 ± 0.04	0.49 ± 0.07	0.48 ± 0.03	0.43 ± 0.04	0.53 ± 0.07	0.54 ± 0.10
	Feather	0.47 ± 0.06	0.46 ± 0.08	0.52 ± 0.03	0.50 ± 0.04	0.49 ± 0.04	0.54 ± 0.08
CCT	FC	0.53	0.56	0.50	0.51	0.56	0.61
	NMF-IBNE	0.46	0.46	0.56	0.53	0.50	0.50
	LDP	0.52	0.54	0.49	0.47	0.50	0.58
	Graph2Vec	0.49	0.50	0.49	0.49	0.47	0.49
	GL2Vec	0.50	0.50	0.50	0.50	0.49	0.50
	DWC	0.57	0.57	0.50	0.50	0.50	0.57
	Scattering	0.57	0.52	0.49	0.49	0.56	0.55
	Feather	0.50	0.51	0.51	0.48	0.57	0.57

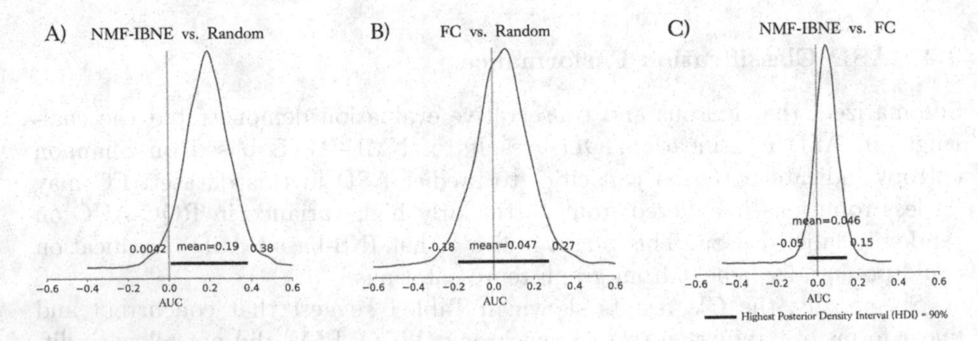

Fig. 1. (A) Strong evidence was found for superior AUC of NMF-IBNE compared to model performance on randomized labels. (B) FC showed high uncertainty and no evidence of predictive validity in the randomization test. (C) Posterior distribution indicated a weak tendency of increased AUC in NMF-IBNE compared to FC.

4 Conclusion

Recent studies in clinical imaging and other areas of machine learning have shown that despite an increasing model complexity, generalization across benchmark datasets may be limited. Therefore, current ASD classification performance using resting-state data may not be sufficient for clinical applications, even when sample size increases. Developing methods that provide a more precise representation of neural data which is better suited to capture the specific aspects of a disorder is of utmost importance.

Here we introduced unsupervised learning of graph representations that explicitly map the neural mechanisms of a core aspect of ASD (i.e. deficits in dyadic social interaction) and combined these with classification algorithms. We employed a rigorous evaluation regime to ensure predictive validity.

These first results indicate that ASD classification based on hyperscanning data is still challenging, yet network embeddings might contribute to improve the development of biomarkers. Furthermore, the choice of the connectivity estimator appears to be important. In particular, nonlinear structures of time-varying synchrony, e.g. captured by Shannon entropy, should be explored in greater detail in hyperscanning. Further, the results indicate that HbO and HbR, which are differentially affected by physiological effects and artifacts, might be treated as related but distinct features in future ASD studies. Certainly, further benchmarking is necessary in larger samples, however, our results indicate that the collection of such datasets should be pushed forward to advance the development and validation of neural biomarkers in ASD. Importantly, our proposed method is suitable for young children, an age at which behavioral diagnostic markers are less reliable but at the same time, it is an early detection and intervention that can dramatically increase long-term outcomes in ASD. Notably, this approach may move forward the scientific investigation of ASD beyond diagnostic classification: Interpretable network embeddings such as NMF-IBNE are trained in a task-agnostic fashion, are thus applicable to many settings and allow for flexible downstream applications, e.g. to address inferential scientific inquiries such as changes after intervention.

References

1. Babiloni, F., Astolfi, L.: Social neuroscience and hyperscanning techniques: past, present and future. Neurosci. Biobehav. Rev. **44**, 76–93 (2014)
2. Baxter, A.J., Brugha, T., Erskine, H.E., Scheurer, R.W., Vos, T., Scott, J.G.: The epidemiology and global burden of autism spectrum disorders. Psychol. Med. **45**(3), 601–613 (2015)
3. Benavoli, A., Corani, G., Demšar, J., Zaffalon, M.: Time for a change: a tutorial for comparing multiple classifiers through Bayesian analysis. J. Mach. Learn. Res. **18**(77), 1–36 (2017)
4. Bolis, D., Schilbach, L.: Observing and participating in social interactions: action perception and action control across the autistic spectrum. Developm. Cognit. Neurosci. **29**, 168–175 (2018)

5. Bouthillier, X., et al.: Accounting for variance in machine learning benchmarks. In: Smola, A., Dimakis, A., Stoica, I. (eds.) Proceedings of Machine Learning and Systems, vol. 3, pp. 747–769 (2021)
6. Brodersen, K.H., et al.: Dissecting psychiatric spectrum disorders by generative embedding. NeuroImage **4**, 98–111 (2014)
7. Cai, C., Wang, Y.: A simple yet effective baseline for non-attributed graph classification. arXiv preprint arXiv:1811.03508 (2018)
8. Chanel, G., Pichon, S., Conty, L., Berthoz, S., Chevallier, C., Grèzes, J.: Classification of autistic individuals and controls using cross-task characterization of fMRI activity. NeuroImage **10**, 78–88 (2016)
9. Chen, H., Koga, H.: GL2vec: graph embedding enriched by line graphs with edge features. Neural Inf. Process. **11955**, 3–14 (2019)
10. Dawson, G., et al.: Randomized, controlled trial of an intervention for toddlers with autism: the Early Start Denver Model. Pediatrics **125**(1), e17–e23 (2010)
11. Ecker, C., Bookheimer, S.Y., Murphy, D.G.: Neuroimaging in autism spectrum disorder: brain structure and function across the lifespan. Lancet Neurol. **14**(11), 1121–1134 (2015)
12. Gao, F., Wolf, G., Hirn, M.: Geometric scattering for graph data analysis. In: Proceedings of the 36th International Conference on Machine Learning, pp. 2122–2131. PMLR (2019). ISSN: 2640–3498
13. Gerloff, C., Konrad, K., Bzdok, D., Büsing, C., Reindl, V.: Interacting brains revisited: a cross-brain network neuroscience perspective. Hum. Brain Mapp. **43**(14), 4458–4474 (2022)
14. Hosseini, M., et al.: I tried a bunch of things: the dangers of unexpected overfitting in classification of brain data. Neurosci. Biobehav. Rev. **119**, 456–467 (2020)
15. Kipf, T.N., Welling, M.: Semi-supervised classification with graph convolutional networks. arXiv preprint arXiv:1609.02907 (2016)
16. Kruppa, J.A., et al.: Brain and motor synchrony in children and adolescents with ASD-a fNIRS hyperscanning study. Soc. Cognit. Affect. Neurosci. **16**(1–2), 103–116 (2021)
17. Narayanan, A., Chandramohan, M., Venkatesan, R., Chen, L., Liu, Y., Jaiswal, S.: graph2vec: Learning Distributed Representations of Graphs. arXiv preprint arXiv:1707.05005 (2017)
18. Pooch, E.H.P., Ballester, P., Barros, R.C.: Can we trust deep learning based diagnosis? The impact of domain shift in chest radiograph classification. In: Petersen, J., et al. (eds.) TIA 2020. LNCS, vol. 12502, pp. 74–83. Springer, Cham (2020). https://doi.org/10.1007/978-3-030-62469-9_7
19. Quiñones-Camacho, L.E., Fishburn, F.A., Belardi, K., Williams, D.L., Huppert, T.J., Perlman, S.B.: Dysfunction in interpersonal neural synchronization as a mechanism for social impairment in autism spectrum disorder. Autism Res. **14**(8), 1585–1596 (2021)
20. American Psychiatric Association: Diagnostic and Statistical Manual of Mental Disorders, 5th edn. American Psychiatric Association, Washington DC (2013)
21. Reindl, V., et al.: Multimodal hyperscanning reveals that synchrony of body and mind are distinct in mother-child dyads. NeuroImage **251**, 118982 (2022)
22. Rozemberczki, B., Sarkar, R.: Characteristic functions on graphs: birds of a feather, from statistical descriptors to parametric models. In: Proceedings of the 29th ACM International Conference on Information & Knowledge Management, pp. 1325–1334. ACM, Virtual Event Ireland (2020)

23. Schulte-Rüther, M., et al.: Using machine learning to improve diagnostic assessment of ASD in the light of specific differential and co-occurring diagnoses. J. Child Psychol. Psychiat. (2022)
24. Scott-Van Zeeland, A.A., Dapretto, M., Ghahremani, D.G., Poldrack, R.A., Bookheimer, S.Y.: Reward processing in autism. Autism Res. **3**(2), 53–67 (2010)
25. Tanabe, H.C., et al.: Hard to "tune in": neural mechanisms of live face-to-face interaction with high-functioning autistic spectrum disorder. Front. Human Neurosci. **6**, 268 (2012)
26. Traut, N., et al.: Insights from an autism imaging biomarker challenge: promises and threats to biomarker discovery. NeuroImage **255**, 119171 (2022)
27. Varoquaux, G., Cheplygina, V.: Machine learning for medical imaging: methodological failures and recommendations for the future. NPJ Digit. Med. **5**(1), 1–8 (2022)
28. Wang, L., Huang, C., Ma, W., Cao, X., Vosoughi, S.: Graph embedding via diffusion-wavelets-based node feature distribution characterization. In: Proceedings of the 30th ACM International Conference on Information & Knowledge Management, pp. 3478–3482. ACM, Queensland (2021)
29. Wang, Q., et al.: Autism symptoms modulate interpersonal neural synchronization in children with autism spectrum disorder in cooperative interactions. Brain Topography **33**(1), 112–122 (2020)
30. Zeidan, J., et al.: Global prevalence of autism: a systematic review update. Autism Res. **15**(5), 778–790 (2022)

fMRI-S4: Learning Short- and Long-Range Dynamic fMRI Dependencies Using 1D Convolutions and State Space Models

Ahmed El-Gazzar[1,2(✉)], Rajat Mani Thomas[1,2], and Guido van Wingen[1,2]

[1] Department of Psychiatry, Amsterdam UMC Location University of Amsterdam, Amsterdam, The Netherlands
gazzar033@gmail.com

[2] Amsterdam Neuroscience, Amsterdam, The Netherlands

Abstract. Single-subject mapping of resting-state brain functional activity to non-imaging phenotypes is a major goal of neuroimaging. The large majority of learning approaches applied today rely either on static representations or on short-term temporal correlations. This is at odds with the nature of brain activity which is dynamic and exhibit both short- and long-range dependencies. Further, new sophisticated deep learning approaches have been developed and validated on single tasks/datasets. The application of these models for the study of a different targets typically require exhaustive hyperparameter search, model engineering and trial and error to obtain competitive results with simpler linear models. This in turn limit their adoption and hinder fair benchmarking in a rapidly developing area of research. To this end, we propose **fMRI-S4**; a versatile deep learning model for the classification of phenotypes and psychiatric disorders from the timecourses of resting-state functional magnetic resonance imaging scans. fMRI-S4 capture short- and long-range temporal dependencies in the signal using 1D convolutions and the recently introduced state-space models **S4**. The proposed architecture is lightweight, sample-efficient and robust across tasks/datasets. We validate fMRI-S4 on the tasks of diagnosing major depressive disorder (MDD), autism spectrum disorder (ASD) and sex classification on three multi-site rs-fMRI datasets. We show that fMRI-S4 can outperform existing methods on all three tasks and can be trained as a *plug&play* model without special hyperpararameter tuning for each setting (Code available at https://github.com/elgazzarr/fMRI-S4.)

Keywords: Functional connectivity · State space models · 1D CNNs · Major depressive disorder · Autism spectrum disorder

1 Introduction

Predicting non-imaging phenotypes from functional brain activity is one of the major objectives of the neuroimaging and neuroscience community. The ability to map a scan of the brain to behaviour or phenotypes would advance our

A. Abdulkadir et al. (Eds.): MLCN 2022, LNCS 13596, pp. 158–168, 2022.
https://doi.org/10.1007/978-3-031-17899-3_16

understating of the brain function and enable the investigation of the underlying pathophysiology of psychiatric disorders. To build such prediction models, researchers have opted for machine learning to capture multivariate patterns in brain functional activity that might act as a bio-marker for the phenotype in question [1,32]. Functional magnetic resonance imaging (fMRI) offers a promising non-invasive tool to estimate brain functional activity by measuring the blood-oxygenation-level-dependent (BOLD) signal as a proxy of the underlying neuronal activity of the brain [23]. From a machine learning point of view, fMRI is one of the most challenging data representations. The high dimensionality of the data (4D, $\approx$ 1M voxels), low signal to noise ratio, data heterogeneity and limited sample sizes present major hurdles when developing learning models from fMRI data. To overcome some of these limitations, researchers have opted for summarized representations that facilitate learning from such data and can be interpreted for biomarker discovery. One of the most popular representations is functional connectivity (FC) [10]. FC is simply defined as the temporal correlation in the BOLD signal changes between different regions of interest in the brain (ROIs). The study of FC have provided a wealth of knowledge about the brain function and dysfunction and have proved that there exists underlying neural correlates of phenotypes that a machine learning model can learn [11]. Yet, FC in its most popular form is a static representation which implicitly assumes that brain connectivity is static for the duration of the scan. In recent years, there has been growing evidence that brain connectivity is dynamic as the brain switch from cognitive states even at rest and more research is currently studying dynamic functional connectivity [18,26]. With the advancement of deep learning, researchers have opted for sequential models such as RNNs, LSTMs, 1D CNNs and Transformers to learn from ROIs timecourses directly [5,7,8,12,22,39]. Because the number of parameters of these models scale with the length of the timecourse and fMRI datasets are typically limited in sample sizes, applying such models to cover the entire duration of the resting-state scan is highly prone to overfitting. Further, recurrent models are not parallelizable and similar to attention-based models are computationally expensive to train. A practical solution is then to either i) crop the ROI timecourses, train the models on the cropped sequences, then aggregate the predictions from all the crops (e.g. via voting) to generate the final prediction at inference [5,8,12]. or ii) limit the effective receptive field of the model using smaller kernels and fewer layers [7]. The former method is prevalent with recurrent and attention based models while the latter is typically observed with convolutional models. fMRI BOLD signals display rich temporal organization, including scale-free $1/f$ power spectra and long-range temporal auto-correlations, with activity at any given time being influenced by the previous history of the system up to several minutes into the past [15,18]. Learning only from short-range temporal interactions (e.g. 30–40 s as typically done when cropping or defining dynamic FC windows) ignores the evolution of the signal and the sequential switching through cognitive states and is more susceptible to psychological noise.

Apart from capturing the true underlying dynamics of the data, a key consideration when developing machine learning models is their utility and adoption by the community. Today, for a researcher interested in investigating a certain phenotype or a clinical outcome using machine learning, the main go-to remains FC analysis using linear models or *shallow* kernel based methods. The main reasons for this are i) **Simplicity**: easy to implement and interpret without much engineering and hyperparameter tuning, ii) **Sample-efficiency**; possible to train with small sample sizes as is often the case in clinical datasets. iii) **Computational-efficiency**; do not require special hardware. iv) **Performance**; most importantly is that they achieve the desired objective. Several recent studies have reported competitive performance of linear models or shallow non-linear models against DL in phenotype prediction from neuroimaging data [16,28].

To bridge this gap and to address the dynamic limitation in several DL architectures, we propose fMRI-S4; a powerful deep learning model that leverages 1D-CNNs and state-space models to learn short- and long-range saptio-temporal features from rs-fMRI data. fMRI-S4 is open-source, data-efficient, easy to train, and can outperform existing methods on phenotype prediction without requiring special hyperparameter tuning for each task/dataset/ROI-template. We validate our work on three multi-site datasets encompassing three different targets. Namely, the ***UkBiobnak*** [30] for sex classification, ***ABIDE*** [4] for autism ASD diagnosis, and ***Rest-Meta-MDD*** [38] for MDD diagnosis.

2 Methodology

2.1 Preliminaries

Phenotype prediction from rs-fMRI data can be formulated as a multivariate timeseries classification problem. Let $X \in R^{N \times T}$ represent a resting-state functional scan where N denotes the number of brain ROIs as extracted using a pre-defined template (*The spatial dimension*), and T represent the number of timepoints sampled for the duration of the scan (*The temporal dimension*). Given a labelled multi-site dataset $D = \{X_i, Y_i\}_{i=1}^{S}$, where Y is a non-imaging phenotype or a disorder diagnosis, the objective is to learn a parameterized function f_θ : $X \mapsto Y$. This is under the practical limitations of small sample size S(typically in the order of hundreds in rs-fMRI multi-site datasets), and that scan duration T and temporal resolution Tr are variable within the dataset since it is collected from different scanning sites. The main challenge then becomes how to design a generalizable f_θ that can capture the underlying causal variables in the data necessary to predict the target.

2.2 Learning Short-Range Dependencies with 1D Convolutions

The resting-state signal is characterized by low-frequency oscillations (0.01–0.1 Hz). To extract informative features from such signal, a model has to learn the short-range dynamic dependencies that characterize the neural activations responsible for generating the oscillations. While in essence, a feature extraction layer that scans the entire global signal (e.g. Transformers, Fully connected layers, RNNs) can learn such local dependencies, leveraging the inductive bias of convolutions significantly improve sample- and parameter- efficiency of the model. Convolutional layers excel at learning local patterns using a translation invariant sliding window and have been used in conjugation with *global* layers to alleviate the memory bottleneck and enhance locality [21,29].

We utilize a convolutional encoder as first stage in the model to learn local dependencies, improve the signal to noise ratio and to mix the features in the spatial dimension. Our encoder consists of K_{conv} blocks, where each blocks consists of a 1D convolutional layer with kernel size k and hidden dimension d_model, followed by batch normalization, and then *relu* activation. 1D CNNs treat the N ROIs as input channels and thus the temporal kernel is fully connected across the spatial dimension. This is a very useful property since the ordering of the ROIs is arbitrary and the model is implicitly free to learn any spatial dependencies necessary for the objective regardless of the parcellation template applied during pre-processing.

2.3 Learning Long-Range Dependencies with State Space Models

1D convolutions are a powerful tool to model spatio-temporal dependencies. Yet, they are limited by their receptive field in the temporal dimension. While this can be improved by adding more layers, increasing the kernel size or using dilation [24], the output features at each layer remain constrained by the receptive field of the respective convolutional filter size. Given that 1DCNN models are typically small (1–3 layers) [7,36], the output features only represent local dependencies. Further, since fMRI datasets are usually collected at multiple locations using different scanners, the temporal resolution of the data may vary across the dataset, which deems finding an optimal kernel size and dilation factors a challenging engineering task. A state space model (SSM) is a linear representation of dynamical systems in continuous or discrete form. The objective of SSMs is to compute the optimal estimate of the hidden state given the observed data [37]. SSMs have been widely in modelling fMRI signal, and its applications includes decoding mental state representation [6,17,19], inferring effective connectivity [33], generative models for classifcation [31]. Inspired by the recent developments in SSMs, specifically the **S4** model [14], which utilizes state-space models as sequence-to-sequence trainable layers and excel at long-range tasks, we propose to integrate S4 layers in our fMRI classifier to capture the global dependencies in the signal for single-subject prediction of non-imaging phenotypes. SSM in continuous time-space maps a 1-D input signal $u(t)$ to an M-D latent state $z(t)$ before projecting to a 1-D output signal $y(t)$ as follows:

$$z^{'}(t) = Az(t) + Bu(t) \tag{1}$$

$$y(t) = Cz(t) + Du(t) \tag{2}$$

where $A \in R^{M \times M}$ is the state-transition matrix, $B \in R^{M \times 1}$, $C \in R^{1 \times M}$ and $D \in R^{1 \times 1}$ are projection matrices. To operate on discrete-time sequences sampled with a step size of Δ, the SSM can be discretized using the bilinear method [34] as follows:

$$z_k = \bar{A} z_{k-1} + \bar{B} u_k \qquad y_k = \bar{C} z_k + \bar{D} u_k \tag{3}$$

$$\bar{A} = (I - \Delta/2 \cdot A)^{-1}(I + \Delta/2 \cdot A) \qquad \bar{B} = (I - \Delta/2 \cdot A)^{-1} \Delta B) \qquad \bar{C} = C \tag{4}$$

Given an initial state $z_k = 0$ and omitting D (as it can be represented as a skip connection in the model), unrolling 3 yields:

$$\begin{aligned} y_k &= \bar{C}\bar{A}^k \bar{B} u_0 + \bar{C}\bar{A}^{k-1}\bar{B} u_1 + ... + \bar{C}\bar{B} u_k \\ y &= \bar{K} * u \qquad \bar{K} = (\bar{C}\bar{A}^i \bar{B})_{i \in [L]} \end{aligned} \tag{5}$$

The operator $\bar{K}$ can thus be interpreted as a convolution filter and the state space model can be trained as a sequence-to-sequence layer via learning parameters A, B, C and Δ with gradient descent. Training $\bar{K}$ efficiently requires several computational tricks. The S4 paper proposes the parameterization of A as a diagonal plus low-rank (DPLR) matrix. This parameterization has two key properties. First, this is a structured representation that allows faster computation using the Cauchy-kernel algorithm [25] to compute the convolution kernel K very quickly. Second, this parameterization includes certain special matrices called HiPPO matrices [13], which theoretically and empirically allow the SSM to capture long-range dependencies better via memorization. For in-depth details of the model we refer the readers to [27]. SSMs defines a map from $R^L \mapsto R^L$, i.e. a 1-D sequence map. To handle multi-dimensional inputs/features $R^{L \times H}$, the S4 layer simply defines H independent copies of itself at and after applying a non-linear activation function and layer normalization, the H feature maps are mixed with a position-wise linear layer. This defines a single S4 block. In our model, we stack K_{s4} blocks on top of the convolutional layers, followed by a global average pooling layer across the temporal dimension, a dropout layer and a Linear layer with output dimension equal to the number of classes. Figure 1 demonstrates the overall architecture of the model.

2.4 fMRI-S4: Towards a Fixed Baseline

A key aspect when designing a classification model for fMRI is versatility. Ideally, we would like to use fixed architecture and fixed training parameters for any dataset/any target and obtain competitive results without the need of exhaustive hyperparameter search and model engineering. This would improve the utility and accessibility of the model for practitioners with different technical backgrounds, and facilitate models benchmarking. The proposed fMRI-S4 model constitutes several desirable properties that makes it feasible to find such optimal

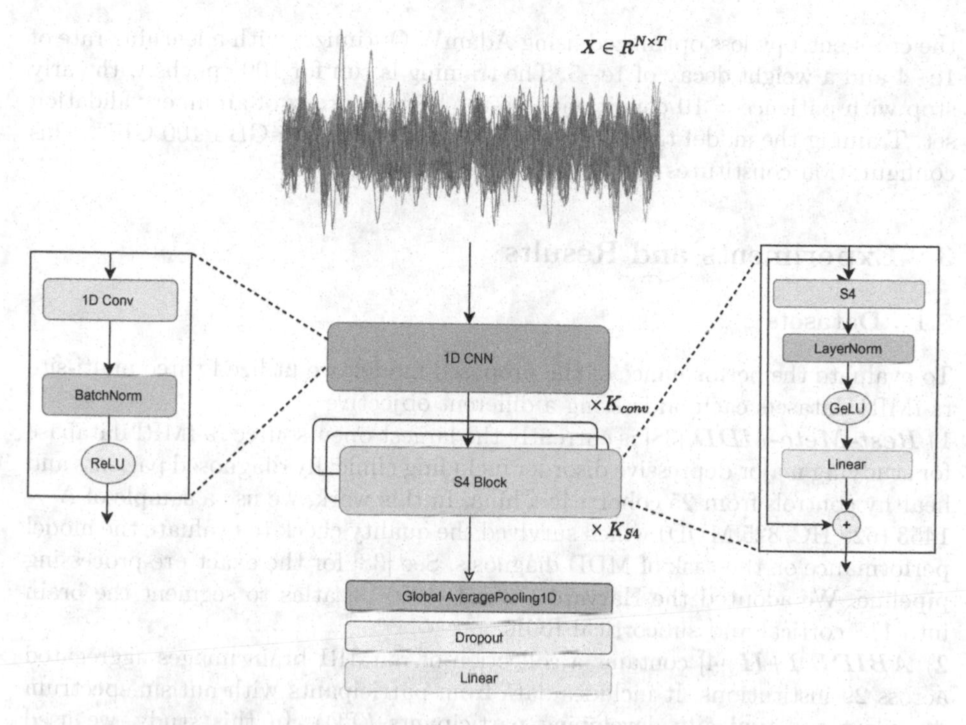

Fig. 1. An overview of the fMRI-S4 Model. The 1D CNN Blocks learn short-range temporal dependencies using a small kernel size and learn spatial dependencies across the ROIs. The output features are then fed to a cascade of S4 Blocks to learn both short- and long-range temporal dependencies.

architecture. The sliding window approach in 1D convolutions enable feature extraction independent of the length of the time course. Further, it's permutation invariant in the spatial dimension, i.e. it can map any number of ROIs with any arbitrary ordering into a fixed dimension. An S4 layer learn a global representation of the data, and can thus eliminate the need to tailor the number of layers in the model to cover the desired receptive field. Moreover, S4 learn an adaptive discritization step size Δ, which further improve the flexibility of the model with respect to the variable temporal resolution of the datasets. This setup in turn can improve the odds of finding a set of optimal parameters for the model and training that can achieve competitive results invariant to the task and the dataset. To this end, we conducted Bayesian hyperparameter tuning using the weights&biases platform [2] on independent validation sets on the three datasets presented in Sect. 3.1 to find the optimal configuration of the model. The top performing configurations converged to a highly overlapping set and based on this we report our default configuration of the model which we use for all the experiments in this work and recommend as a baseline. For the model architecture we use $d_{model} = 256$, $K_{conv} = 1$, $K_{S4} = 2$, $d_{state} = 256$. The rest of the parameters are fixed as in the original S4 model. We trained the models using the cross-entropy loss optimized using AdamW Optimizer with a learning rate of 1e−4 and a weight decay of 1e−5. The training is run for 100 epochs with early stop with patience = 10 conditioned on the best accuracy of an inner-validation set. Training the model takes ≈ 15–20 min on a Nividia 16-GB P100 GPU. This configuration constitutes 1.3 M trainable parameters.

3 Experiments and Results

3.1 Datasets

To evaluate the performance of the proposed model, we utilized three multi-site rs-fMRI datasets each addressing a different objective.
1) ***Rest-Meta-MDD*** [38] is currently the largest open-source rs-fMRI database for studying major depressive disorder including clinically diagnosed patients and healthy controls from 25 cohorts in China. In this work, we use a sample of $N = 1453$ (628 HC/825 MDD) which survived the quality check to evaluate the model performance on the task of MDD diagnosis. See [38] for the exact pre-processing pipeline. We adopted the Harvard Oxford (HO) [3] atlas to segment the brain into 118 cortical and subcortical ROIs.
2) ***ABIDE I+II*** [4] contains a collection of rs-fMRI brain images aggregated across 29 institutions. It includes data from participants with autism spectrum disorders and typically developing participants (TD). In this study, we used a subset of the dataset with $N = 1207$ (558 TD/649 ASD) to evaluate the model performance on the tasks of ASD diagnosis. We utilized the C-PAC pre-processing pipeline to pre-process the data. Similarly, we adopted the HO atlas to segment the brain.
3) ***UkBioBank*** [30] is a large-scale population database, containing in-depth genetic and health information from half a million UK participants. In this work we use a randomly sampled subset (N = 5500 (2750 M/2750 F F) to evaluate the model performance on the sex classification task. The ROIs were extracted using the Automated Anatomical Labeling (AAL) [35] atlas.

3.2 Clinical Results

We evaluated the performance of the fMRI-S4 model against following baselines: **SVM**, **BrainNetCNN** [20], **1D-CNN** [7], **ST-GCN** [12] and **DAST-GCN** [9]. The input representation to the **SVM** and **BrainNetCNN** are the static Person correlation matrices of the ROIs timecourses, while the rest of the methods operate directly on the ROIs timecourses. For a fair evaluation, we conduct a hyperparameter search using weights&biases [2] for the baselines on a independent validation set to select the best configuration for each task. Next, we conducted the experiments using a repeated 5-fold cross validation scheme on the two clinical tasks using the selected parameters for the baselines and the fixed configuration presented in Sect. 2.4 for fMRI-S4. We report the results in Table 1. For both tasks, fMRI-S4 outperform the best performing baseline (the

1D-CNN) by 1.6 accuracy points (2.5% relative) and 2.4 accuracy points (3.4% relative) on the ***Rest-Meta-MDD*** and ***ABIDE I+II*** datasets, respectively. To demonstrate the efficacy of combining both 1D convolutions and state space models, we conduct a simple ablation study with two models, i) fMRI-S4$_{K_{S4}=0}$; where the S4 layers are removed and replaced by 2 convolutional layers. ii) fMRI-S4$_{K_{conv}=0}$; where no convolution layers are used, and spatio-temporal feature extraction is done using S4 layers only. In both cases, the accuracy of the model drops by 2–4% for both datasets.

Table 1. 5-fold test metrics for fMRI-S4, ablated version of fMRI-S4 and baseline models on the two clinical datasets.

Model	*Rest-Meta-MDD*			*ABIDE I+II*		
	Acc. (%)	Sens. (%)	Spec. (%)	Acc. (%)	Sens. (%)	Spec. (%)
SVM	61.9 ± 3	73.0 ± 9	50.9 ± 8	69.5 ± 3	75.9 ± 5	63.0 ± 5
BrainNetCNN [20]	58.4 ± 5	50.1 ± 11	56.8 ± 8	66.6 ± 3	72.1 ± 4	61.2 ± 5
ST-GCN [12]	58.2 ± 4	48.6 ± 9	67.8 ± 7	65.3 ± 2	67.2 ± 3	63.4 ± 4
DAST-GCN [9]	60.7 ± 4	44.6 ± 7	75.8 ± 7	67.8 ± 2	70.8 ± 3	64.9 ± 3
1D-CNN [7]	63.8 ± 2	65.7 ± 3	61.9 ± 3	70.6 ± 2	72.7 ± 3	68.5 ± 3
fMRI-S4$_{K_{S4}=0}$	63.7 ± 3	68.5 ± 6	59.7 ± 6	71.7 ± 2	75.7 ± 5	67.8 ± 3
fMRI-S4$_{K_{conv}=0}$	62.9 ± 2	67.5 ± 5	60.1 ± 5	70.4 ± 3	74.5 ± 3	66.1 ± 1
fMRI-S4	$\mathbf{65.4 \pm 3}$	66.9 ± 4	63.9 ± 4	$\mathbf{73.0 \pm 3}$	75.2 ± 4	70.8 ± 3

4 Evaluating Sample-Efficiency with the UkBioBnak

To compare the sample efficiency of fMRI-S4 against existing baseline, we conduct a training sample scaling experiment on the UkbioBank dataset for the task of sex classification. Namely, we train the models using N = [500, 1000, 2000, 5000] class-balanced samples, and evaluate the performance of the trained models on a fixed test set with N = 500 (250 M/250 F). The results presented on Fig. 2 highlight that fMRI-S4 performs competitively at the smallest sample sizes (N = 500) and continue to scale favourably against the baselines. Another observation from the results is the scaling superiority of the dynamic models (1D-CNNs, fMRI-S4, DAST-GCN) over static models with exception of the ST-GCN model. This behaviour suggests that the evolution of the dynamic signal contain discriminate information that can be better exploited by the deep learning models at larger training samples.

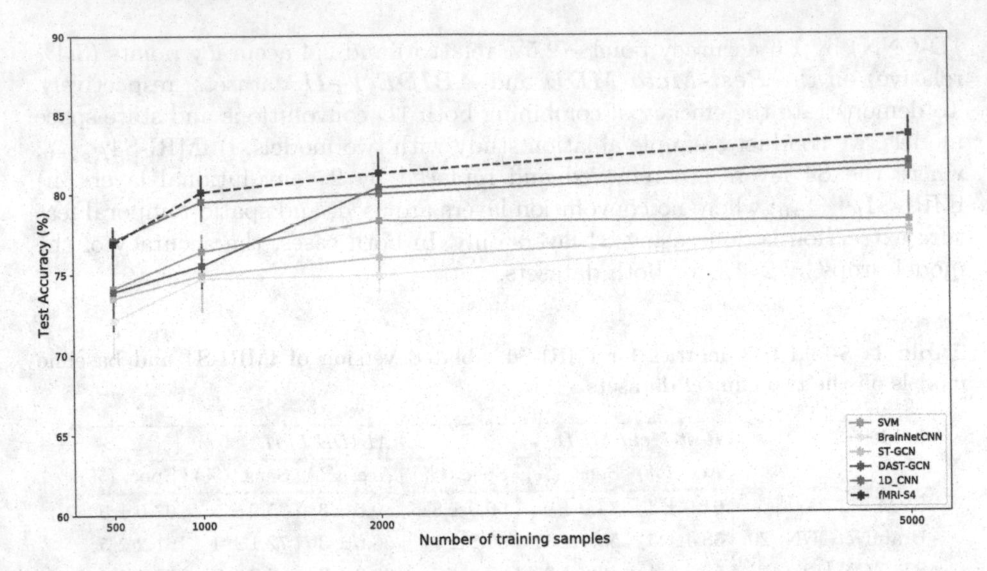

Fig. 2. Scaling performance of the fMRI-S4 against existing baselines on the task of sex classification on the UkbioBank dataset. The error bars represent results for 3 different random seeds for initialisation and inner validation split. The test set is fixed for all the experiments.

5 Discussion

In this work, we present fMRI-S4; a deep learning model that leverages 1D convolutions and state-space models for learning short- and long-range dependencies necessary to capture the underlying dynamic evolution of the brain activity at rest. We show fMRI-S4 improves the diagnosis of MDD, ASD and sex classification from rs-fMRI data against existing methods using a fixed architecture for all three tasks. We hope that this work can improve the adoption of dynamic DL-based models in fMRI analysis and motivate the development of generalizable methods. In our future work, we will investigate dynamic feature perturbation to explain the predictions of fMRI-S4 in effort to obtain potential biomarkers for psychiatric disorders.

References

1. Arbabshirani, M.R., Plis, S., Sui, J., Calhoun, V.D.: Single subject prediction of brain disorders in neuroimaging: promises and pitfalls. Neuroimage **145**, 137–165 (2017)
2. Biewald, L.: Experiment tracking with weights and biases (2020). https://www.wandb.com/. Software available from wandb.com
3. Desikan, R.S., et al.: An automated labeling system for subdividing the human cerebral cortex on MRI scans into gyral based regions of interest. Neuroimage **31**(3), 968–980 (2006)

4. Di Martino, A., et al.: The autism brain imaging data exchange: towards a large-scale evaluation of the intrinsic brain architecture in autism. Mol. Psychiatry **19**(6), 659–667 (2014)
5. Dvornek, N.C., Ventola, P., Pelphrey, K.A., Duncan, J.S.: Identifying autism from resting-state fMRI using long short-term memory networks. In: Wang, Q., Shi, Y., Suk, H.-I., Suzuki, K. (eds.) MLMI 2017. LNCS, vol. 10541, pp. 362–370. Springer, Cham (2017). https://doi.org/10.1007/978-3-319-67389-9_42
6. Eavani, H., Satterthwaite, T.D., Gur, R.E., Gur, R.C., Davatzikos, C.: Unsupervised learning of functional network dynamics in resting state fMRI. In: Gee, J.C., Joshi, S., Pohl, K.M., Wells, W.M., Zöllei, L. (eds.) IPMI 2013. LNCS, vol. 7917, pp. 426–437. Springer, Heidelberg (2013). https://doi.org/10.1007/978-3-642-38868-2_36
7. El Gazzar, A., Cerliani, L., van Wingen, G., Thomas, R.M.: Simple 1-d convolutional networks for resting-state fMRI based classification in autism. In: 2019 International Joint Conference on Neural Networks (IJCNN), pp. 1–6. IEEE (2019)
8. El-Gazzar, A., Quaak, M., Cerliani, L., Bloem, P., van Wingen, G., Mani Thomas, R.: A hybrid 3DCNN and 3DC-LSTM based model for 4d spatio-temporal fMRI data: an ABIDE autism classification study. In: Zhou, L., et al. (eds.) OR 2.0/MLCN -2019. LNCS, vol. 11796, pp. 95–102. Springer, Cham (2019). https://doi.org/10.1007/978-3-030-32695-1_11
9. El-Gazzar, A., Thomas, R.M., van Wingen, G.: Dynamic adaptive spatio-temporal graph convolution for fMRI modelling. In: Abdulkadir, A., et al. (eds.) MLCN 2021. LNCS, vol. 13001, pp. 125–134. Springer, Cham (2021). https://doi.org/10.1007/978-3-030-87586-2_13
10. Friston, K.J.: Functional and effective connectivity in neuroimaging: a synthesis. Hum. Brain Mapp. **2**(1–2), 56–78 (1994)
11. Friston, K.J.: Functional and effective connectivity: a review. Brain Connect. **1**(1), 13–36 (2011)
12. Gadgil, S., Zhao, Q., Pfefferbaum, A., Sullivan, E.V., Adeli, E., Pohl, K.M.: Spatio-temporal graph convolution for resting-state fMRI analysis. In: Martel, A.L., et al. (eds.) MICCAI 2020. LNCS, vol. 12267, pp. 528–538. Springer, Cham (2020). https://doi.org/10.1007/978-3-030-59728-3_52
13. Gu, A., Dao, T., Ermon, S., Rudra, A., Ré, C.: HiPPO: recurrent memory with optimal polynomial projections. Adv. Neural. Inf. Process. Syst. **33**, 1474–1487 (2020)
14. Gu, A., Goel, K., Ré, C.: Efficiently modeling long sequences with structured state spaces. In: International Conference on Learning Representations (2022)
15. He, B.J.: Scale-free properties of the functional magnetic resonance imaging signal during rest and task. J. Neurosci. **31**(39), 13786–13795 (2011)
16. He, T., et al.: Deep neural networks and kernel regression achieve comparable accuracies for functional connectivity prediction of behavior and demographics. Neuroimage **206**, 116276 (2020)
17. Hutchinson, R.A., Niculescu, R.S., Keller, T.A., Rustandi, I., Mitchell, T.M.: Modeling fMRI data generated by overlapping cognitive processes with unknown onsets using hidden process models. Neuroimage **46**(1), 87–104 (2009)
18. Hutchison, R.M., et al.: Dynamic functional connectivity: promise, issues, and interpretations. Neuroimage **80**, 360–378 (2013)
19. Janoos, F., Machiraju, R., Singh, S., Morocz, I.A.: Spatio-temporal models of mental processes from fMRI. Neuroimage **57**(2), 362–377 (2011)
20. Kawahara, J., et al.: BrainNetCNN: convolutional neural networks for brain networks; towards predicting neurodevelopment. Neuroimage **146**, 1038–1049 (2017)

21. Li, S., et al.: Enhancing the locality and breaking the memory bottleneck of transformer on time series forecasting. Advances in Neural Information Processing Systems 32 (2019)
22. Malkiel, I., Rosenman, G., Wolf, L., Hendler, T.: Pre-training and fine-tuning transformers for fMRI prediction tasks. arXiv preprint arXiv:2112.05761 (2021)
23. Ogawa, S., Lee, T.M., Kay, A.R., Tank, D.W.: Brain magnetic resonance imaging with contrast dependent on blood oxygenation. Proc. Natl. Acad. Sci. **87**(24), 9868–9872 (1990)
24. Oord, A.v.d., et al.: WaveNet: a generative model for raw audio. arXiv preprint arXiv:1609.03499 (2016)
25. Pan, V.: Fast approximate computations with Cauchy matrices and polynomials. Math. Comput. **86**(308), 2799–2826 (2017)
26. Preti, M.G., Bolton, T.A., Van De Ville, D.: The dynamic functional connectome: state-of-the-art and perspectives. Neuroimage **160**, 41–54 (2017)
27. Sasha, R., Sidd, K.: The annotated s4. In: Blog Track at ICLR 2022 (2022). https://srush.github.io/annotated-s4/
28. Schulz, M.A., et al.: Different scaling of linear models and deep learning in UKBiobank brain images versus machine-learning datasets. Nat. Commun. **11**(1), 1–15 (2020)
29. Shi, X., Chen, Z., Wang, H., Yeung, D.Y., Wong, W.K., Woo, W.C.: Convolutional LSTM network: a machine learning approach for precipitation nowcasting. Advances in Neural Information Processing Systems 28 (2015)
30. Sudlow, C., et al.: UK biobank: an open access resource for identifying the causes of a wide range of complex diseases of middle and old age. PLoS Med. **12**(3), e1001779 (2015)
31. Suk, H.I., Wee, C.Y., Lee, S.W., Shen, D.: State-space model with deep learning for functional dynamics estimation in resting-state fMRI. Neuroimage **129**, 292–307 (2016)
32. Sundermann, B., Herr, D., Schwindt, W., Pfleiderer, B.: Multivariate classification of blood oxygen level-dependent fMRI data with diagnostic intention: a clinical perspective. Am. J. Neuroradiol. **35**(5), 848–855 (2014)
33. Tu, T., Paisley, J., Haufe, S., Sajda, P.: A state-space model for inferring effective connectivity of latent neural dynamics from simultaneous EEG/fMRI. Advances in Neural Information Processing Systems **32** (2019)
34. Tustin, A.: A method of analysing the behaviour of linear systems in terms of time series. J. Inst. Electr. Eng. Part IIA Autom. Regul. Servo Mech. **94**(1), 130–142 (1947)
35. Tzourio-Mazoyer, N., et al.: Automated anatomical labeling of activations in SPM using a macroscopic anatomical parcellation of the MNI MRI single-subject brain. Neuroimage **15**(1), 273–289 (2002)
36. Wang, Z., Yan, W., Oates, T.: Time series classification from scratch with deep neural networks: a strong baseline. In: 2017 International Joint Conference on Neural Networks (IJCNN), pp. 1578–1585. IEEE (2017)
37. Williams, R.L., Lawrence, D.A., et al.: Linear State-Space Control Systems. Wiley, Hoboken (2007)
38. Yan, C.G., et al.: Reduced default mode network functional connectivity in patients with recurrent major depressive disorder. Proc. Natl. Acad. Sci. **116**(18), 9078–9083 (2019)
39. Yan, W., et al.: Discriminating schizophrenia using recurrent neural network applied on time courses of multi-site fMRI data. EBioMedicine **47**, 543–552 (2019)

Data Augmentation via Partial Nonlinear Registration for Brain-Age Prediction

Marc-Andre Schulz[1(✉)], Alexander Koch[1], Vanessa Emanuela Guarino[2], Dagmar Kainmueller[2], and Kerstin Ritter[1]

[1] Department of Psychiatry and Neurosciences, Charité - Universitätsmedizin Berlin, Corporate Member of Freie Universität Berlin and Humboldt-Universität zu Berlin, Berlin, Germany
marc.schulz@rwth-aachen.de
[2] Max-Delbrueck-Center for Molecular Medicine in the Helmholtz Association (MDC), Berlin, Germany

Abstract. Data augmentation techniques that improve the classification and segmentation of natural scenes often do not transfer well to brain imaging data. The conceptually most plausible augmentation technique for biological tissue, elastic deformation, works well on microscopic tissue but is limited on macroscopic structures like the brain, as the majority of mathematically possible elastic deformations of the human brain are anatomically implausible. Here, we characterize the subspace of anatomically plausible deformations for a participant's brain image by nonlinearly registering the image to the brain images of several reference participants. Using the resulting warp fields for data augmentation outperformed both random elastic deformations and the non-augmented baseline in age prediction from T1 brain images.

Keywords: Brain imaging · Machine learning · Data augmentation

1 Introduction

Digital processing of human brain images promises automated diagnosis and prediction of treatment responses as well as insight into the underlying neurobiology of neurological and psychiatric disease. Many such use cases rely on a-priori unknown high-dimensional patterns in imaging data which may be invisible to the human eye, so researchers turn to machine learning to identify predictive statistical relationships in the data [1,2].

However, modern machine learning models often require larger amounts of training data than are customarily collected in human brain imaging studies [3]. Particularly, clinical imaging datasets are often limited to small sample sizes

M.-A. Schulz and A. Koch—Equal contribution. Project was funded by DFG 414984028/CRC-1404) and the Brain & Behavior Research Foundation (NARSAD young investigator grant).

A. Abdulkadir et al. (Eds.): MLCN 2022, LNCS 13596, pp. 169–178, 2022.
https://doi.org/10.1007/978-3-031-17899-3_17

due to small patient pools as well as logistical and financial constraints. In cases where a data-hungry machine learning model needs to be trained on a small-sample dataset, researchers often rely on data augmentation techniques [4]. The idea behind data augmentation is to exploit dimensions of variation in the data which are irrelevant for the prediction task at hand to generate additional training samples. For example, in computer vision, it is irrelevant for the classification of cats versus dogs where the particular animal is located in the image, how it is rotated, or what lightning conditions are captured in the image. Given this insight, one may generate additional training samples by applying e.g. affine transformations or intensity transformations to the images.

While these standard augmentation techniques work well in computer vision, and particularly in the classification and segmentation of natural scenes, they do not appear to transfer trivially to brain imaging data [5]. In contrast to natural scenes, where translation, rotation and scale cannot be trivially accounted for during a preprocessing stage, brain image acquisition and the resulting images are highly homogeneous, and sophisticated software already takes care of skull stripping and linear and nonlinear registration to a standard space. Critically, these preprocessing steps are done by default to enable classical analyses such as voxel-based morphometry and are further thought to reduce prediction-irrelevant variance in the data so that little is gained by skipping these steps for machine learning [6]. In sum, while there are obvious invariances in natural scenes (e.g. lighting conditions might be different depending on the time of day), there are (in the opinion of the authors) no obvious invariances for pre-registered brain imaging data (e.g. no time-of-day effect on the intensity values in brain images). Thus, most augmentation techniques used in computer vision do not successfully transfer to neuroimaging-based phenotype prediction - in contrast to e.g. brain tumor segmentation, which is more of a traditional computer vision task. For instance, Dufumier et al. [5], report generally detrimental effects of affine transformations, intensity transformations, and mirroring on prediction accuracy for a variety of prediction tasks on structural neuroimaging data.

One potentially plausible augmentation is elastic deformation [7–9]. Elastic deformations appear to work well for segmentation tasks, both on microscopy images, e.g. for segmenting cells [8], as well for lesion segmentation on macroscopic human tissue [10,11]. The underlying intuition is that human tissue is elastic - "squishy" - and thus can be elastically deformed - yielding substantially more degrees of freedom than simple affine transformations. These intuitions should hold for brain images. However, while there are reports of successful use of elastic deformation for brain tumor segmentation [10,12], there are, to the best of our knowledge, no literature reports of positive results for classification/regression tasks.

We conjecture that, while on a microscopic level most deformations make sense, macroscopic deformations on the whole brain are qualitatively different. Anatomic heterogeneity in the brain happens on a limited set of dimensions of variation. The space of possible elastic deformations is infinite so that in the absence of a generative model of the brain, it is unclear which possible whole-brain elastic deformations are anatomically meaningful. Implausible deformations may degrade training performance by obscuring relevant information and degrade generalisation performance by overtaxing the model's expressive capacity. Consequently, a major challenge in the application of elastic deformation augmentation in neuroimaging is characterizing the potentially tiny subspace of anatomically plausible elastic deformations.

In this paper, we describe a method for characterizing the subspace of biologically plausible deformations by borrowing deformation fields from nonlinear image registration. We (partially) register a given image to the brain image of a different participant as a form of data augmentation. Essentially, we ask: How would this brain look like in another person's skull? We experimentally validate the proposed data augmentation by partial registration on age prediction from structural brain images.

Related Work. Other authors have applied similar approaches to anatomical image segmentation [13,14]. However, it is unclear if results from image segmentation will transfer to classification/regression tasks. Moreover, neither study implemented a random elastic deformation baseline so that it is unclear whether their custom deformations outperformed augmentation via random elastic deformations.

2 Data Augmentation via Partial Nonlinear Registration

In the following sections, we introduce the fundamentals of elastic deformations, identify the shared mathematical basis with nonlinear image registration, and then show how to exploit these commonalities for data augmentation of brain images.

Elastic Deformations. Elastic deformations can be generated directly on the voxel level using a random displacement field convolved with a Gaussian filter [7] or generated using a free-form deformation by modifying an underlying mesh of control points [8,9]. The field of displacement vectors for the control points is interpolated, often using cubic B-splines [15,16], into a voxel-level deformation field or "warp field". For data augmentation, one generally samples and applies a random warp field each time one wants to augment the image, thus generating potentially infinite randomly deformed versions of the original. Incidentally, free-form deformation using warp fields represented as cubic B splines are also used for nonlinear image registration, e.g. for registering a T1 image to the nonlinear MNI template (see the following section) using FSL-FNIRT [17].

Biologically (Im-)plausible Elastic Deformations. Elastic deformations allow for considerable degrees of freedom that result in essentially infinite potential warp fields. A majority of potential warp fields are anatomically implausible (c.f. Fig. 1). How to meaningfully limit the space of possible elastic deformations to anatomically plausible solutions is a challenging problem. Given the underlying shared mathematics, we suggest using warp fields derived from nonlinear

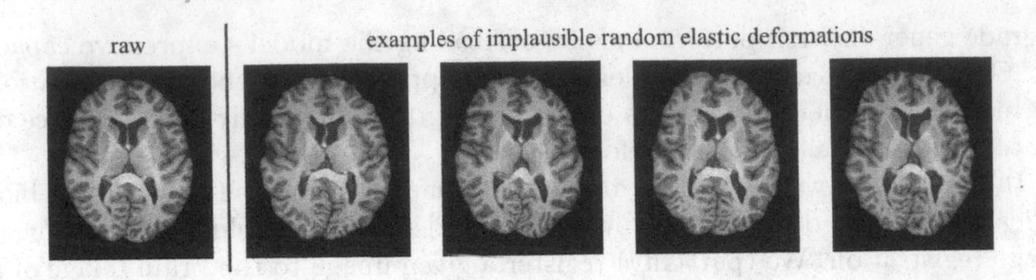

Fig. 1. Randomly samples deformation fields can yield anatomically implausible results.

registration to multiple reference images to span a subspace of plausible elastic deformations.

Nonlinear Registration to a Reference Image is an Elastic Deformation. To compare MRI images of different subjects, e.g. for voxel-based morphometry, one needs to relate corresponding voxels between images - a process called image registration. The goal of image registration, is to relate any source point $(x, y, z) \in \mathbb{Z}^3$ to a point in the reference image (x', y', z'), i.e. a transform $T : (x, y, z) \mapsto (x', y', z')$ needs to be found. This is generally achieved by a combined transformation, consisting of a global and a local transformation [18].

In the simplest case, the global transformation can be described as a rigid body transformation, with six degrees of freedom, describing the translation and rotation of the image. More generally, global transformations can be expressed as affine transformations, offering six additional degrees of freedom, allowing to express scaling and shearing. This approach is generally referred to as linear registration.

Applying only a global linear transformation can be insufficient. Additional local nonlinear transformations can be used to improve the registration quality [17]. Such local transformations are typically parameterized as free-form deformations. Free-form deformations allow the manipulation of the source using a mesh of control points that, when manipulated, modifies the shape of the underlying image. Effectively, nonlinear registration is an elastic deformation that relies on the same underlying framework as augmentation via random elastic deformations.

Partial Nonlinear Registration. We define a partial registration as a linearly interpolated registration between a linearly registered image and a nonlinearly registered one. Using a fixed interpolation factor $\lambda \in [0, 1]$, we can blend between both registrations, creating a partially registered image. Setting λ to zero we obtain a linearly registered image, while setting λ to one, we obtain a fully nonlinearly registered image. This technique allows us to randomly sample a λ for partial registration to create multiple augmented images from a given brain image, representing different degrees of registration to a given reference image.

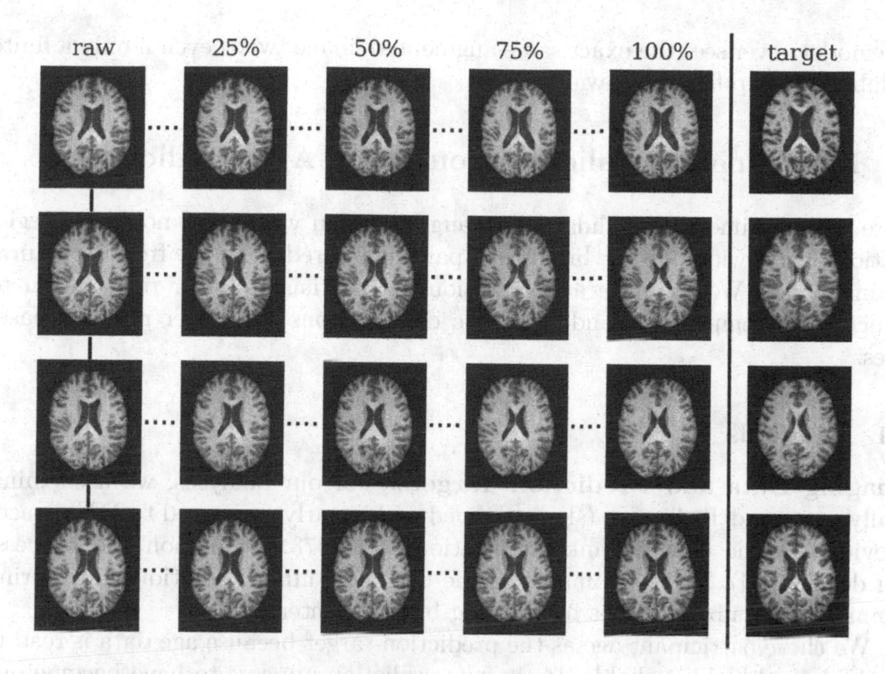

Fig. 2. For data augmentation, we nonlinearly register the source image to images from several reference participants and retain the resulting deformation fields. At train time, we randomly sample a reference participant, re-scale the corresponding deformation field by a random factor, and use these re-scaled fields to warp the original image.

Registration to Multiple Reference Subjects. Conventionally images are only co-registered between different imaging modalities of the same subject, or they are registered to a standard space such as MNI space. However, images can also be registered onto the brain of another person. Qualitatively, this procedure is comparable to registering to a standard space MNI template.

Application for Data Augmentation. In aggregate, we suggest the following approach for data augmentation: In the preprocessing phase, each training image is registered to reference images of several different participants, discarding the output image but retaining the resulting warp fields. For machine learning, every time a given participant's brain image is used in a training step, we randomly select one of the pre-computed warp fields for that participant and then sample a random percentage for partial registration. In different training steps, participant A's brain image may be used practically unchanged (approx. 0% partial registration) or slightly warped to the brain and skull shape of participant B (e.g. 10% partial registration) or strongly warped to the brain and skull shape of participant C (e.g. 90% partial registration). See Fig. 2 for an illustration. This should ensure that the warp fields used for elastic deformation data augmentation represent variation observable in the real world, i.e. anatomical variation between different participants. Moreover, the partial registration approach ensures that the model never sees the exact same augmented image twice, even if only a finite number of warp fields are available.

3 Experimental Validation on Brain-Age Prediction

Here we experimentally validate data augmentation via partial nonlinear registration on the widely-used "brain-age" paradigm, predicting age from structural brain images. We compare augmentation via partial nonlinear registration to structurally comparable random elastic deformations and other relevant baselines.

3.1 Methods

Imaging Data and Prediction Targets. For our analyses, we used minimally processed T1 images (skull stripped and linearly registered to MNI space) provided by the UK Biobank (application no. 33073, acquisition and processing details [19]). We used images in the original 1 mm^3 resolution and, during training, normalized images by dividing by mean intensity.

We chose participant age as the prediction target because age data is readily available and highly reliable. Brain-age prediction appears to have become one of the most popular research paradigms in the application of machine learning on neuroimaging data [20,21]. Ages were z-scored for training.

Warp Fields. We randomly selected 50 participants from the UK Biobank to serve as anatomical reference images. The T1 image of each training participant was nonlinearly registered to each of the reference participant's T1 images, and warp fields were retained for data augmentation. We used FSL-FNIRT for nonlinear registration, using the default configuration (10 mm control point grid), except for a simplified sub-sampling schedule (4, 2, 1) for faster computation [17]).

Deep Neural Network Architecture and Training Regime. We roughly replicated the deep learning setup for brain-age prediction of Fish et al. [22], thus minimizing our own design choices. Specifically, we used a ResNet-50 architecture in the 3D variant by Hara [23]. The Adam optimizer was applied with a 1cycle learning rate schedule [24] with 0.01 maximum learning rate over 50000 gradient update steps, minimizing mean squared error as a loss function. Other hyperparameters are left as defaults provided by the PyTorch implementation. No additional form of regularisation, such as weight decay or dropout was used.

Evaluation. We randomly selected 1000 participants from the UK Biobank as a train set and a further 500 participants as a test set. No form of hyperparameter tuning took place within this study so that there was no need for an extra validation set. Each experiment was repeated for three random seeds.

We compare our approach to three relevant baselines: The *non-augmented baseline* applies no changes to the training data and represents a majority of described training setups in the literature [25–27]. *Mirroring* i.e. flipping on the sagittal plane, yielded promising results in our own preliminary experiments and is used e.g. in [28,29]. *Random elastic deformations* use the same 10mm control point grid as returned by FNIRT, but sample displacement vectors randomly from a uniform distribution. Per convention, we limit displacements to half the distance between control points to avoid folding (reducing the maximal displacement by a factor of 0.5 or 0.1 yielded inferior results in preliminary experiments). This configuration is meant to ensure comparability between random elastic deformations and deformations derived from partial nonlinear registration.

3.2 Results

After training, we observed a test set mean absolute error (MEAN $\pm$ SD in years) of 3.61 ± 0.08 for partial registration, 4.10 ± 0.03 for random elastic deformations, 4.10 ± 0.12 for mirroring, and 4.23 ± 0.05 for the non-augmented baseline. The same rank order (in reverse) was observed in the training loss, with the non-augmented baseline achieving the lowest training loss, followed by mirroring, followed by the random elastic deformations, and tailed by partial registrations with the highest training loss. In sum, our method outperformed the strongest baseline by a 3.4σ margin. However, the overall variance in results between different random seeds was substantial, so the results should be interpreted with caution (Fig 3).

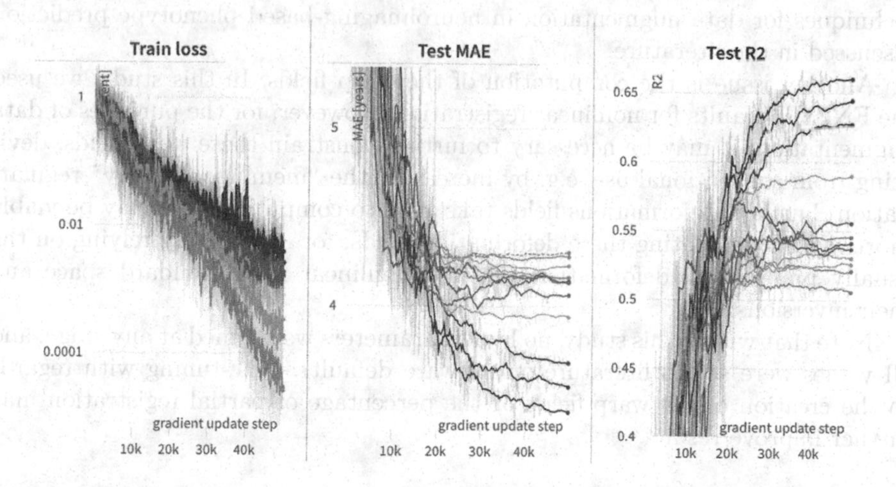

Fig. 3. Train and test set metrics for age prediction from 1000 minimally preprocessed T1 images using no augmentation, mirroring, random elastic deformations, the proposed partial nonlinear registration. Moving window average smoothed. (Color figure online)

4 Discussion

In this work, we discuss the lack of effective and biologically plausible data augmentation techniques for brain imaging data. Due to the elasticity of biological tissue, we identify elastic deformations as a promising candidate technique. To limit the infinite space of possible elastic deformations, we take advantage of deformation fields derived from nonlinear registration to a reference image. On the widely-used "brain-age" paradigm, we compared augmentation by partial nonlinear registration to structurally comparable random elastic deformations and other relevant baselines. We observed substantial improvements in prediction accuracy using partial nonlinear registration. The increase in accuracy when moving from random elastic deformations to augmentation by partial nonlinear registration indicates that the latter did indeed provide superior, potentially more anatomically plausible, augmentations.

While our results appear promising, at this point, there remain a number of limitations. First of all, we can make no claim of generality for different types of input data, different resolutions of input data, different preprocessing regimes, nor can we claim generality with regards to different target variables. It is yet unclear how effective augmentation techniques are at different sample sizes using different deep learning architectures. Finally, it is unclear to what extent a full (100%) registration to a reference image makes sense, given that it would, for example, substantially alter the size of the brain ventricles, which are an important predictor for age and other target phenotypes. Partial registration may have to be limited to a lower maximal warping percentage for optimal results. Please note that these limitations are not unique to this work but apply to practically all techniques for data augmentation in neuroimaging-based phenotype prediction discussed in the literature.

Another issue is the computation of the warp fields. In this study, we used the FNIRT defaults for nonlinear registration. However, for the purposes of data augmentation, it may be necessary to further constrain these warp fields, deviating from conventional use, e.g. by increasing the "membrane energy" regularisation. Further, deformations fields take time to compute. There may be viable shortcuts for generating these deformation fields, for example by relying on the usually pre-existing deformation fields to nonlinear MNI standard space and their inversions.

Note that within this study, no hyperparameters were tuned at any stage, and all values were set to literature or software defaults. Fine-tuning with regards to the creation of the warp fields or the percentage of partial registration may further improve results.

References

1. Bzdok, D., Meyer-Lindenberg, A.: Machine learning for precision psychiatry: opportunities and challenges. Biol. Psychiat. Cogn. Neurosci. Neuroimag. **3**, 223–230 (2018)

2. Eitel, F., Schulz, M.-A., Seiler, M., Walter, H., Ritter, K.: Promises and pitfalls of deep neural networks in neuroimaging-based psychiatric research. Exp. Neurol. **339**, 113608 (2021)
3. Schulz, M.-A., Bzdok, D., Haufe, S., Haynes, J.-D., Ritter, K.: Performance reserves in brain-imaging-based phenotype prediction, bioRxiv (2022)
4. Shorten, C., Khoshgoftaar, T.M.: A survey on image data augmentation for deep learning. J. Big Data **6**, 1–48 (2019)
5. Dufumier, B., Gori, P., Battaglia, I., Victor, J., Grigis, A., Duchesnay, E.: Benchmarking CNN on 3D anatomical brain MRI: architectures, data augmentation and deep ensemble learning, arXiv:2106.01132 [cs, eess], June 2021. arXiv: 2106.01132
6. Klingenberg, M., Stark, D., Eitel, F., Ritter, K.: MRI image registration considerably improves CNN-based disease classification. In: Abdulkadir, A., et al. (eds.) MLCN 2021. LNCS, vol. 13001, pp. 44–52. Springer, Cham (2021). https://doi.org/10.1007/978-3-030-87586-2_5
7. Simard, P.Y., Steinkraus, D., Platt, J.C.: Best practices for convolutional neural networks applied to visual document analysis. In: 7th International Conference on Document Analysis and Recognition (ICDAR 2003), 2-Volume Set, Edinburgh, Scotland, UK, 3–6 August 2003, pp. 958–962. IEEE Computer Society (2003)
8. Ronneberger, O., Fischer, P., Brox, T.: U-Net: convolutional networks for biomedical image segmentation. In: Navab, N., Hornegger, J., Wells, W.M., Frangi, A.F. (eds.) MICCAI 2015. LNCS, vol. 9351, pp. 234–241. Springer, Cham (2015). https://doi.org/10.1007/978-3-319-24574-4_28
9. Çiçek, Ö., Abdulkadir, A., Lienkamp, S.S., Brox, T., Ronneberger, O.: 3D U-Net: learning dense volumetric segmentation from sparse annotation. In: Ourselin, S., Joskowicz, L., Sabuncu, M.R., Unal, G., Wells, W. (eds.) MICCAI 2016. LNCS, vol. 9901, pp. 424–432. Springer, Cham (2016). https://doi.org/10.1007/978-3-319-46723-8_49
10. Cirillo, M.D., Abramian, D., Eklund, A.: What is the best data augmentation for 3d brain tumor segmentation?. In: 2021 IEEE International Conference on Image Processing (ICIP), pp. 36–40. IEEE (2021)
11. Castro, E., Cardoso, J.S., Pereira, J.C.: Elastic deformations for data augmentation in breast cancer mass detection. In: 2018 IEEE EMBS International Conference on Biomedical & Health Informatics (BHI), pp. 230–234. IEEE (2018)
12. Wang, G., Li, W., Ourselin, S., Vercauteren, T.: Automatic brain tumor segmentation based on cascaded convolutional neural networks with uncertainty estimation. Front. Comput. Neurosci. **13**, 56 (2019)
13. Shen, Z., Xu, Z., Olut, S., Niethammer, M.: Anatomical data augmentation via fluid-based image registration. In: Martel, A.L., et al. (eds.) MICCAI 2020. LNCS, vol. 12263, pp. 318–328. Springer, Cham (2020). https://doi.org/10.1007/978-3-030-59716-0_31
14. Nalepa, J., et al.: Data augmentation via image registration. In: 2019 IEEE International Conference on Image Processing (ICIP), pp. 4250–4254. IEEE (2019)
15. Tustison, N.J., Avants, B.B., Gee, J.C.: Directly manipulated free-form deformation image registration. IEEE Trans. Image Process. **18**(3), 624–635 (2009)
16. Gu, S., et al.: Bidirectional elastic image registration using B-spline affine transformation. Comput. Medical Imaging Graph. **38**(4), 306–314 (2014)
17. Andersson, J.L.R., Jenkinson, M., Smith, S.M.: Non-linear registration aka spatial normalisation (2007)
18. Rueckert, D., Sonoda, L., Hayes, C., Hill, D., Leach, M., Hawkes, D.: Nonrigid registration using free-form deformations: application to breast MR images. IEEE Trans. Med. Imaging **18**(8), 712–721 (1999)

19. Alfaro-Almagro, F., et al.: Image processing and Quality Control for the first 10,000 brain imaging datasets from UK Biobank. Neuroimage **166**, 400–424 (2018)
20. Cole, J.H., et al.: Predicting brain age with deep learning from raw imaging data results in a reliable and heritable biomarker. Neuroimage **163**, 115–124 (2017)
21. Kaufmann, T., et al.: Common brain disorders are associated with heritable patterns of apparent aging of the brain. Nat. Neurosci. **22**(10), 1617–1623 (2019)
22. Fisch, L., et al.: Predicting brain-age from raw t1-weighted magnetic resonance imaging data using 3d convolutional neural networks (2021)
23. Hara, K., Kataoka, H., Satoh, Y.: Can spatiotemporal 3D CNNs retrace the history of 2D CNNs and ImageNet? In: 2018 IEEE Conference on Computer Vision and Pattern Recognition, pp. 6546–6555. Computer Vision Foundation/IEEE Computer Society (2018)
24. Smith, L.N., Topin, N.: Super-convergence: very fast training of neural networks using large learning rates (2017)
25. Jiang, H., et al.: Predicting brain age of healthy adults based on structural MRI parcellation using convolutional neural networks. Front. Neurol. **10**, 1346 (2020)
26. Baecker, L., et al.: Brain age prediction: a comparison between machine learning models using region- and voxel-based morphometric data. Hum. Brain Mapp. **42**(8), 2332–2346 (2021)
27. Popescu, S.G., Glocker, B., Sharp, D.J., Cole, J.H.: Local brain-age: a U-Net model. Front. Aging Neurosci. **13** (2021)
28. Peng, H., Gong, W., Beckmann, C.F., Vedaldi, A., Smith, S.M.: Accurate brain age prediction with lightweight deep neural networks. Med. Image Anal. **68**, 101871 (2021)
29. Bashyam, V.M., et al.: MRI signatures of brain age and disease over the lifespan based on a deep brain network and 14 468 individuals worldwide. Brain **143**, 2312–2324 (2020)

Author Index

Printed in the United States
by Baker & Taylor Publisher Services